Seismicity Caused by Mines, Fluid Injections, Reservoirs, and Oil Extraction

Edited by
Shahriar Talebi

Springer Basel AG

Reprint from Pageoph
(PAGEOPH), Volume 153 (1998), No. 1

The Editor:

Prof. Dr.-Ing.
Shahriar Talebi
CANMET
555 Booth St., BCC 10
Ottawa, Ontario
Canada K1A 0G1

A CIP catalogue record for this book is available from the Library of Congress, Washington D.C., USA

Deutsche Bibliothek Cataloging-in-Publication Data

Seismicity caused by mines, fluid injections, reservoirs, and oil extraction /
ed. by Shahriar Talebi. - Basel ; Boston ; Berlin : Birkhäuser 1999
   (Pageoph topical volumes)
    ISBN 978-3-7643-6048-1   ISBN 978-3-0348-8804-2 (eBook)
    DOI 10.1007/978-3-0348-8804-2

© 1999 Springer Basel AG
Originally published by Birkhäuser Verlag in 1999

Printed on acid-free paper produced from chlorine-free pulp

ISBN 978-3-7643-6048-1

9 8 7 6 5 4 3 2 1

Contents

Pure appl. geophys. 153 (1998) 1–2
0033–4553/98/010001–02 $ 1.50 + 0.20/0

❙Pure and Applied Geophysics

Introduction

This special issue of Pure and Applied Geophysics contains 12 papers dealing with different aspects of induced and triggered seismicity. These papers divide naturally into four groups focusing on seismicity associated with mining, water injections, reservoirs and acquifers, and oil extraction. In many respects the present issue can be regarded as a natural extension of a previous special issue of Pure and Applied Geophysics (Vol. 150, nos. 3/4, 1997).

The first group of four papers investigates mine-induced seismicity. The purpose of the paper by Gibowicz is the study of stress release mode for a large collection of seismic events recorded in different mining environments. Lasocki and Idziak use a technique based on the non-parametric kernel estimation method to analyze the dominant trends of epicenter migration of regional seismicity. Alcott et al. discuss the application of a methodology based on the use of seismic source parameters to assess rockburst hazard in mines. Cai et al. propose a tensile failure model that seems to provide realistic source dimensions for microseismic events around an underground opening. The intent here is to emphasize the role of the fracture surface energy in the failure process that, in reality, entails other modes of failure as well. The second group of 2 papers describes comprehensive results of monitoring and analysis of microseismicity induced by water injection in an oil field.

The third group of four papers focuses on seismicity triggered by reservoirs and acquifers. Chen and Talwani review case histories of this phenomenon in China and show that the triggering processes are different for granitic and karst terranes. Muço presents the results of twenty years of monitoring induced seismicity in northern Albania. Mandal and Rastogi estimate the coda Q_c of the Koyna-Warna region. Bella et al. study the seismicity induced by one of the largest acquifers of central Italy. The fourth group of two papers concerns seismicity due to steam stimulation and oil extraction. The first paper presents a seismic model of casing failures in oil fields and the second paper summarizes the results of monitoring seismic activity in an oil field over a 2-year period.

I would like to sincerely thank Renata Dmowska, editor-in-chief for topical issues at Pure and Applied Geophysics, for inviting me to serve as a guest editor, handling the review process of my own papers, and providing invaluable advice during the preparation of this special issue. Many thanks to Brian Mitchell and the authors of three papers of the present collection, originally submitted to a regular issue of Pure and Applied Geophysics, for having agreed with the inclusion of these papers in this volume. Thanks are also due to the management of Mining and Mineral Sciences Laboratories, CANMET, for providing support. The completion of this volume was made possible thanks to the time and efforts of many scientists who performed conscientious reviews. In that regard, I cordially thank Art McGarr and George Gibowicz for their assistance with the review process and their continued support. Also, I gratefully acknowledge the contributions of François H. Cornet, Jon B. Fletcher, Abraham Hofstetter, Rick Kry, Charles A. Langston, Brian Mitchell, Stephane Nechtschein, Yang Qingyuan, David W. Simpson, Arthur Snoke, Pradeep Talwani, Cezar I. Trifu, Ted. I. Urbancic, Luc Vandamme and Yizhang Zhong.

Shahriar Talebi
CANMET-MSSL
555 Booth St., BCC 10
Ottawa, Ontario
Canada K1A 0G1

Seismicity Induced by Mining

Pure appl. geophys. 153 (1998) 5–20
0033–4553/98/010005–16 $ 1.50 + 0.20/0

▌Pure and Applied Geophysics

Partial Stress Drop and Frictional Overshoot Mechanism of Seismic Events Induced by Mining

S. J. GIBOWICZ [1]

Abstract—The values of uniformly estimated apparent stress σ_a, and Brune's stress drop, taken as a measure of static stress drop, from 850 seismic events, with moment magnitude ranging from -3.6 to 3.6, induced at the Underground Research Laboratory (URL) in Canada, Western Deep Levels (WDL) gold mine in South Africa, and two coal and two copper mines in Poland, were collected to study the stress release mode in various mining environments. For this, the quantity epsilon, $\varepsilon = \Delta\sigma/(\sigma_a + \Delta\sigma/2)$, where $\Delta\sigma$ is the static stress drop, proposed by ZÚÑIGA (1993) as an indicator of stress-drop mechanism was used. The events induced at the URL are characterized by low values of epsilon corresponding to a partial stress drop mechanism, whereas all the events at WDL display a frictional overshoot mechanism in which final stress reaches a lower value than that of frictional stress. The events at Polish coal and copper mines, on the other hand, are in good agreement with the well-known Orowan's condition such that the final stress is equal to the dynamic frictional stress.

The Brune stress drop, however, is heavily model dependent through the source radius-corner frequency relation. The Orowan's condition for the events from the URL would be met if a constant in the source radius-corner frequency relation is equal to 1.82 ± 0.12, and for the events from WDL if it is equal to 3.92 ± 0.40, in contrast to Brune's constant of 2.34.

The smoothed values of epsilon displayed as a function of time, represented by the consecutive event numbers, for selected sets of events imply that the largest seismic events in a given set occur when the epsilon is low and a partial stress drop mechanism is dominant. The large events are then followed by high epsilon values when a frictional overshoot mechanism begins to dominate.

Key words: Induced seismicity, Orowan's condition, partial stress drop, frictional overshoot, Brune's stress drop, apparent stress.

Introduction

In the well-known simple fault model with constant friction of OROWAN (1960), the final stress after the earthquake rupture is equal to the dynamic frictional stress. Two other possibilities have also been considered. In the first case the final stress is greater than the frictional stress and a partial stress drop mechanism is involved (BRUNE, 1970, 1976). In the second case the final stress is smaller than the frictional

---

[1] Institute of Geophysics, Polish Academy of Sciences, ul. Ks. Janusza 64, 01-452 Warsaw, Poland. E-mail: gibowicz@igf.edu.pl.

stress and a frictional overshoot mechanism is considered (SAVAGE and WOOD, 1971). Several arguments in favor of both cases have been used by various authors (e.g., SAVAGE and WOOD, 1971; BRUNE, 1976; MADARIAGA, 1976; SNOKE *et al.*, 1983; SMITH *et al.*, 1991; ZÚÑIGA, 1993; CASTRO *et al.*, 1997).

Recently, ZÚÑIGA (1993) proposed the use of the quantity epsilon (ε), defined as the ratio of the static stress drop over the sum of apparent stress and half the static stress drop, to study the stress drop mechanism of earthquakes for a simplified fault model, assuming that both static stress drop and apparent stress are known. This ratio is bounded by the values from 0 to 2. The OROWAN'S condition is met when $\varepsilon = 1$, the partial stress drop mechanism corresponds to $\varepsilon < 1$, and the frictional overshoot case appears when $\varepsilon > 1$.

The aim of this study is to examine the stress drop mechanism of seismic events induced by mining using the epsilon of ZÚÑIGA (1993). For this the values of uniformly estimated apparent stress and BRUNE (1970, 1971) stress drop, taken as a measure of static stress drop from 850 seismic events with moment magnitude from −3.6 to 3.6, induced at the Underground Research Laboratory (URL) in Canada, Western Deep Levels (WDL) gold mine in South Africa, Wujek (WJK) and Ziemowit (ZMT) coal mines in Poland, and Polkowice (PLK) and Rudna (RDN) copper mines in Poland were collected (see Table 1). It should be noted that source parameters of seismic events in mines are estimated from records of underground seismic networks where recorded wave forms are not practically affected by free-surface effects, and where attenuation and site effects are expected to be insubstantial.

The static stress drop is defined as the initial stress minus the final stress. It can be expressed in terms of the seismic moment and a characteristic dimension of the fault zone. The seismic moment is not model dependent and can be determined reliably from the low-frequency level of the far-field displacement spectrum. For large earthquakes, the spatial distribution of their aftershocks provides an estimate of the size and shape of the rupture area, leading to a fairly reliable estimate of the

Table 1

Selected mines, number of events, magnitude range, and results of analysis

Mine	Code	Number of events	Magnitude range	Epsilon ratio	Constant K for source size determ.
Underground Res. Lab., Canada	URL	155	− 3.6 to − 1.8	0.674 ± 0.102	1.82 ± 0.12
Western Deep Levels, S. Africa	WDL	199	− 0.5 to 3.1	1.630 ± 0.277	3.92 ± 0.40
Wujek coal mine	WJK	135	0.7 to 2.2	0.980 ± 0.188	2.30 ± 0.17
Ziemowit coal mine	ZMT	47	0.7 to 2.2	0.895 ± 0.173	2.17 ± 0.15
Polkowice copper mine	PLK	172	1.2 to 3.0	1.003 ± 0.176	2.34 ± 0.11
Rudna copper mine	RDN	142	1.3 to 3.6	0.996 ± 0.184	2.33 ± 0.20

static stress drop averaged over the entire fault plane (e.g., BOATWRIGHT, 1984). This is not the case for small seismic events, such as those generated in mines, where only a very few events would cause underground damage whose area could provide an estimate of the rupture surface.

The most commonly used estimate of stress drop, $\Delta\sigma$, follows the source model of BRUNE (1970, 1971)

$$\Delta\sigma = \frac{7}{16}\frac{M_0}{r_0^3},$$ (1)

where M_0 is the seismic moment and r_0 is the radius of a fault, represented by a circular dislocation with instantaneous stress release, expressed as

$$r_0 = (K\beta_0)/(2\pi f_0),$$ (2)

in which K is a constant depending on the source model, for the BRUNE model $K = 2.34$, β_0 is the S-wave velocity in the source area, and f_0 is the corner frequency of S waves.

The constant K in relation (2) takes different values in different source models (BRUNE, 1970; SATO and HIRASAWA, 1973; MADARIAGA, 1976). In mine seismicity studies in South Africa the source size estimated from BRUNE's model has received independent confirmation (e.g., McGARR, 1984, 1991, 1994). In other mining districts, however, the circular fault model of MADARIAGA (1976) has provided better estimates of the source size in good agreement with independent observations in mines (e.g., GIBOWICZ, 1984; GIBOWICZ et al., 1990, 1991; REVALOR et al., 1990).

If the stress drop of BRUNE (1970, 1971), therefore, is accepted as a measure of the static stress drop for all selected events from various mining areas, the procedure used in numerous studies of small seismic events, then the results of our analysis should show the differences in mining environments in terms of different stress release mechanisms there. The other approach is to estimate the values of constant K (relation (2)) in various mining districts corresponding to the OROWAN conjecture. Both approaches will be considered.

The apparent stress, σ_a, defined as the ratio of the radiated energy, E, over the seismic moment, M_0, multiplied by the medium rigidity, μ (WYSS, 1970)

$$\sigma_a = \frac{\mu E}{M_0},$$ (3)

is less model dependent than the static stress drop determined from the seismic moment and corner frequency (SNOKE et al., 1983). If the P-wave contribution to the seismic energy and the azimuth dependence of the energy flux, however, are neglected, then the BRUNE stress drop is a constant multiple of the apparent stress (SNOKE, 1987). Numerous studies of the focal mechanism of seismic events in mines, based on moment tensor inversion (e.g., SATO and FUJII, 1989; FEIGNIER and YOUNG, 1992; McGARR, 1992a,b; WIEJACZ, 1992, 1995), have shown that for a large percentage of these events their source mechanism is more complex than simple shearing represented by a double couple. The energy of P waves could occasionally reach as much as one third of the total seismic energy (e.g., GIBOWICZ, 1996) and cannot be neglected in energy calculations. Our seismic energy, used to estimate the apparent stress (relation (3)), is the sum of P-wave and S-wave energy calculated from the energy flux of both waves. The apparent stress, therefore, becomes an independent parameter (e.g., GIBOWICZ et al., 1991).

Data

The Underground Research Laboratory (URL) is situated within the Lac du Bonnet granite batholith in southeastern Manitoba. A seismic network was installed there to monitor seismicity induced by shaft sinking between 324 and 443 m of depth. The excavation of a 4.6 m diameter circular shaft, performed applying a full-face drill-and-blast technique, was monitored from January to August 1988 by a number of seismic sensors installed in four inclined boreholes drilled from the 300 m level. After each blast a period of increased seismic activity was observed, lasting for about 2 hours. The dynamic range of the system was 92 dB with a resolution of 12 bits; the sampling frequency was 40 kHz.

Source parameters of 155 events with moment magnitude from -3.6 to -1.8 were estimated and listed by GIBOWICZ et al. (1991). The corner frequency, an important parameter in determination of the static stress drop, was calculated from the low-frequency level of displacement spectra and the energy flux of S waves (SNOKE, 1987). The values of apparent stress were taken directly from GIBOWICZ et al. (1991), while the values of stress drop were recalculated by replacing a constant in the source dimension-corner frequency relation from the source model of MADARIAGA (1976), used by GIBOWICZ et al. (1991), by a constant from BRUNE's (1970; 1971) model, to preserve a uniform set of data from seismic events induced in various mining districts.

Western Deep Levels (WDL) gold mine, the deepest mine in the world, is situated in the Carletonville gold mining district. The mine area, extending 11 km from east to west and 4 km from north to south, contains two major conglomerate formations, the Ventersdorp Contact Reef at an average depth of 2 km and the Carbon Leader Reef at a depth of 3 km. On April 7, 1993, a swarm-like sequence of seismic events occurred in the Upper Carbon Leader Back Area at a depth of 3 km. The sequence continued until April 19 and 199 seismic events were recorded with moment magnitude from -0.5 to 3.1 (GIBOWICZ, 1997b). The Integrated Seismic System (ISS), described by MENDECKI (1997), has been in operation at WDL from the beginning of 1990. In 1993 the underground seismic network was composed of 22 three-component stations; several of them in close vicinity to the area where the sequence occurred. The dynamic range of the system is greater than 120 dB with a resolution of 12 bits; the sampling frequency is 2000 Hz. The sequence of April 1993 was recorded, located and processed by the ISS system. The source parameters of seismic events forming the sequence were calculated in an interactive mode by the author.

Two selected coal mines in the Upper Silesian Coal Basin (USCB) in southern Poland are located in different tectonic environments. Wujek (WJK) coal mine is located in the central part of the USCB, on the southern side of the main anticline which is the area of the highest seismicity in the USCB. The mine was founded in 1899 and its mining area is about 8 km^2, containing Quaternary deposits overlaying Carboniferous rocks. The depth of mining ranges from 560 to 740 m. Seismicity at WJK mine is monitored by an underground digital seismic network composed of 14 vertical seismometers, favorably covering several longwalls worked out in 1994 and 1995. The seismic signals are digitized with the sampling frequency of 250 Hz. A 10-bit converter is used and the dynamic range is 60 dB. Ground velocity is recorded and the frequency response is flat between 0.6 and 25 Hz. Seismic events associated with two longwalls were chosen for detailed study. These are longwall 3 in seam 501 at a depth of 670 m, where 70 selected events occurred between January 1994 and February 1995, and longwall 13 in seam 416 at a depth of 650 m, where 65 selected events occurred between July 1994 and April 1995. The moment magnitude range of these events is from 0.7 to 2.2.

Ziemowit (ZMT) coal mine is situated in the southeastern part of the USCB in the main syncline structure. It was founded in 1952 and is one of the biggest coal mines in Poland with an area of about 60 km^2, containing Quaternary, Tertiary, Triassic and Carboniferous deposits. The present depth of mining ranges from 260 to 620 m. The mine is characterized by a middle level of seismicity. Its underground seismic network is composed of 16 vertical seismometers located in the southern part of the mine where the highest seismicity is observed. The seismic signals are digitized with a sampling frequency of 200 Hz. The dynamic range is 68 dB with a resolution of 12 bits. The frequency response is flat between 0.6 and 65 Hz. For analysis 47 seismic events were selected, which occurred between March and July 1994 and were associated with longwall 418 in seam 207. Their moment magnitude ranges from 0.7 to 2.2.

Polkowice (PLK) copper mine is situated in the Lubin mining district between Lubin and Rudna mines in southwestern Poland. The copper ore area is composed of Rotliegendes deposits overlying a crystalline basement. They underly Zechstein deposits of considerable strength. Carbonate deposits, composing the floor of this rigid complex, form the roof of mining works. The carbonate deposits are able to accumulate large strains and then release them suddenly. The single mined seam is composed of dolomite carbonates and dolomites. Seismicity is monitored by a local seismic network composed of 23 vertical and 3 three-component seismometers located at a depth from 460 to 930 m and extending over an area of 10 by 13 km.

The seismic signals are digitized with a sampling frequency of 500 Hz; a 14-bit converter is used. The frequency response is flat from 0.5 to 150 Hz, and the dynamic range is 70 dB (e.g., GIBOWICZ, 1997a). The records of 89 events from mine sections G-21 and G-32, which occurred between May 1994 and May 1995, and of 83 events from sections G-31 and G-51, which occurred between April and August 1996, were selected for analysis. Their moment magnitude ranges from 1.2 to 3.0.

Seismicity at the Rudna (RDN) copper mine in the Lubin mining district is monitored by a seismic network which was composed of 15 vertical seismometers at the close of 1994 and was enlarged to 31 vertical sensors in 1995, located underground at a depth from 550 to 1150 m and extending over an area of approximately 10 by 10 km. The other network characteristics are the same as those of the network at PLK mine. Several sets of seismic events were analyzed for a number of research projects. Altogether, source parameters of 142 events were determined: 28 events from mine sections G-8, G-12, and G-21, which occurred between July and December 1994, 1 large event of May 26, 1995 from section G-23; 82 events from sections G-6, G-12, G-15, and G-21, which occurred between July and December 1995; and 31 events from sections G-1/7 and G-13/3, which occurred between April and December 1996. The range of moment magnitude of these events is from 1.3 to 3.6.

The values of source parameters of selected seismic events from all four Polish mines are listed in a number of local reports and they are available from the author on request. It should be noted that all source parameters of seismic events from various mines, including URL and WDL, used in this study were estimated in a uniform manner from spectral analysis of P and S waves. The procedure is described in detail by GIBOWICZ and KIJKO (1994).

First Approach: Partial Stress Drop and Frictional Overshoot

The relation between apparent stress and stress drop, for a simple fault model with constant friction, can be written as (SAVAGE and WOOD, 1971):

$$\sigma_a = \eta\sigma = \Delta\sigma/2 - (\sigma_f - \sigma_2), \tag{4}$$

where η is the seismic efficiency, $\sigma = (\sigma_1 + \sigma_2)/2$ is the average stress and σ_1 and σ_2 are initial and final stresses, $\Delta\sigma = \sigma_1 - \sigma_2$ is the static stress drop, and σ_f is the frictional stress. For the OROWAN (1960) condition we have:

$$\sigma_a = \Delta\sigma/2. \tag{5}$$

This relation is shown by a straight line in Figure 1 where the values of apparent stress versus stress drop for seismic events from the URL, WDL, and Polish coal and copper mines are presented. From Figure 1 it follows that all the events from

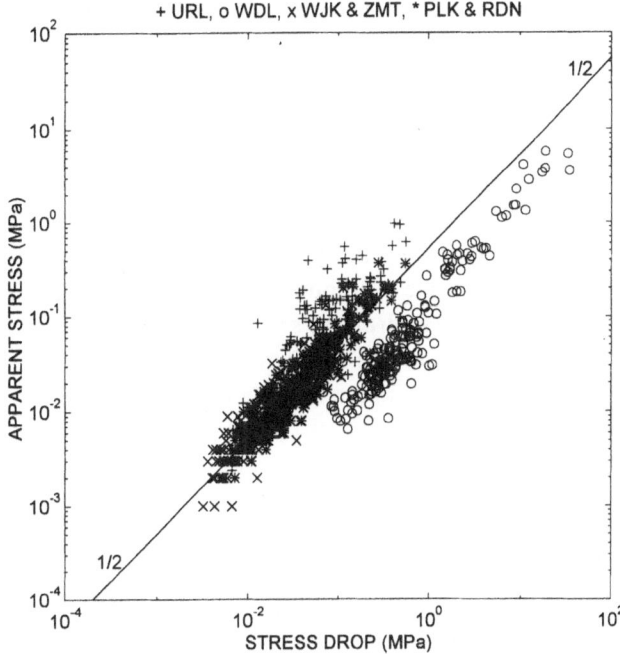

Figure 1

Apparent stress versus Brune's stress drop for seismic events from the Underground Research Laboratory (URL) in Canada, Western Deep Levels (WDL) in South Africa, two Polish coal mines (WJK and ZMT), and two Polish copper mines (PLK and RDN). The straight line indicates the Orowan's condition corresponding to apparent stress equal to half the static stress drop.

WDL clearly display a frictional overshoot mechanism; the events from the URL indicate a partial stress drop mechanism, and all the events from Polish mines fulfill the OROWAN condition, within an acceptable uncertainty resulting from the scatter of data.

To present such a state in a more distinct manner, the quantity ε of ZÚÑIGA (1993) can be used, defined as:

$$\varepsilon = \frac{\Delta\sigma}{\eta\sigma + (\Delta\sigma/2)} = \frac{\sigma_1 - \sigma_2}{\sigma_1 - \sigma_f}. \tag{6}$$

The values of epsilon versus moment magnitude for seismic events from six selected mines are presented in Figure 2, where the OROWAN condition ($\varepsilon = 1$) is marked by a continuous line. Although the scatter of the epsilon, caused by uncertainties in stress parameter estimates and possibly by variable conditions in the source areas, is substantial, it is clear that the stress drop mechanism is not dependent on event magnitude.

Assuming that the scatter in ε values is caused mostly by the uncertainties in σ_a and $\Delta\sigma$ estimates, the mean values of ε and their standard errors for each set of data from the six mines were calculated. For this the standard errors of the apparent stress and stress drop corresponding to their mean values and the law of propagation of errors were taken into account. The results are presented in Figure 3 and are listed in Table 1. The circumstance again is rather clear: the WDL events are characterized by high ε value, including its standard error bounds, well above the OROWAN condition; the URL events show low ε value smaller than one, and the events from Polish mines are characterized by ε values very close to one.

This result was verified by taking into account the dependence of stress parameters on event magnitude. Although constant stress drop scaling relations, implying independence of stress drop on seismic moment, were globally confirmed for seismic events induced by mining within a wide range of seismic moment values (GIBOWICZ, 1995), individual small sets of data often display such a dependence on seismic moment (e.g., MCGARR, 1994). The trade-off between stress drop and rupture velocity inherent in all kinematic source models implies that such a dependence can be attributed just as well to systematic variations of rupture velocity (DEICHMANN, 1997).

The distribution of stress drop values versus moment magnitude for seismic events from the six mines, presented in Figure 4, suggests that such a dependence is apparent. A similar distribution is also observed when apparent stress versus magnitude is plotted. Our six sets of data therefore were divided into 21 sets with

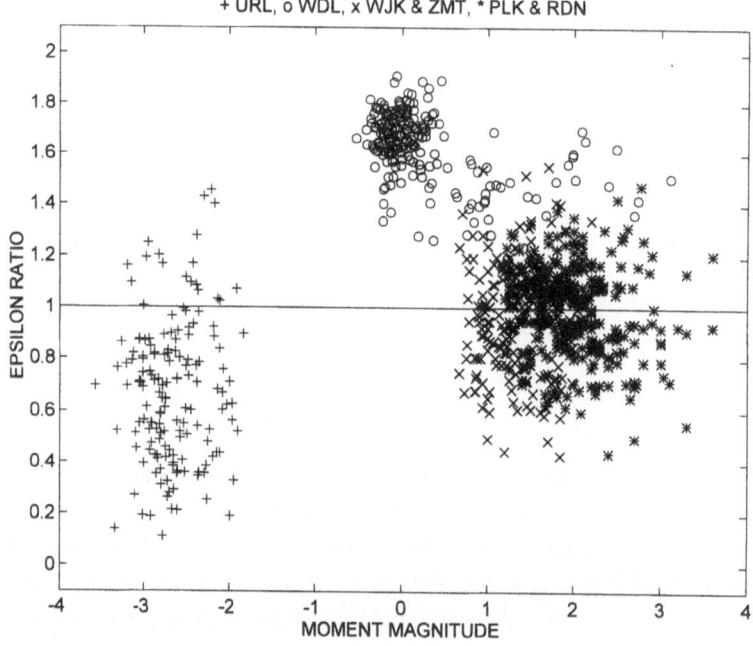

Figure 2

Epsilon ratio versus moment magnitude, based on stress parameter estimates for seismic events from six selected mines. The Orowan's condition corresponds to the epsilon ratio equal to one.

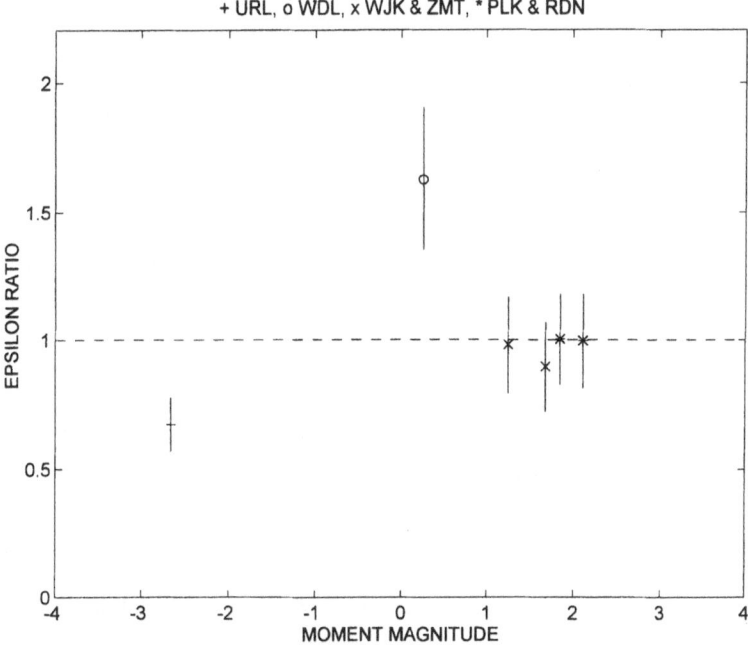

Figure 3

Mean values of the epsilon ratio versus moment magnitude for seismic events from six selected mines; their standard errors are indicated by vertical bars.

magnitude intervals equal approximately to half a unit of magnitude. The epsilon was then calculated for each set and the results are shown in Figure 5. They are similar to those presented in Figure 3 and they confirm again that the seismic events at WDL display frictional overshoot mechanisms; the events at the URL show partial stress drop mechanisms, and the events at Polish mines are in good agreement with the OROWAN condition.

Second Approach: Orowan's Condition and Brune's Stress Drop

If the BRUNE stress drop is not accepted as a measure of the static stress drop, then the alternate approach to the evaluation of differences in stress release mechanism in various mining areas is to estimate the values of a constant K in relation (2), which would replace BRUNE's constant and would provide the stress drop for our selected events from six mines corresponding to the OROWAN condition described by relation (5). We assume therefore that

$$C\Delta\sigma_B = 2\sigma_a, \tag{7}$$

where C is a constant and $\Delta\sigma_B$ is the BRUNE stress drop. The values of a constant C and their standard errors were calculated by least-squares technique from the six sets of data, after which the values of a constant K and their standard errors were determined. The results are listed in Table 1.

The value of $K = 1.82$ obtained from stress release parameters of seismic events at the URL is the middle value between the average value of $K = 1.32$ corresponding to the source model of MADARIAGA (1976) and that of BRUNE (1970, 1971). The coefficient K in the MADARIAGA model depends on the rupture velocity and the azimuth of observation, and the obtained value of $K = 1.82$ could probably be explained in these terms, thus excluding partial stress drop as a possible mechanism for seismic events at the URL. The value of $K = 3.92$ found for the events at WDL, however, seems to be too high for any reasonable explanations in terms of either fault and observation geometry or rupture velocity or both, implying that the frictional overshoot mechanism there is probably real. The values of K found for the events at Polish mines are, of course, close to that of BRUNE.

Possible Temporal Changes in Epsilon

Four sets of data related to seismic events forming well-defined sequences in space and time were selected to consider possible temporal trends in epsilon within these sequences. These are: all 155 events from URL and 199 events from the WDL

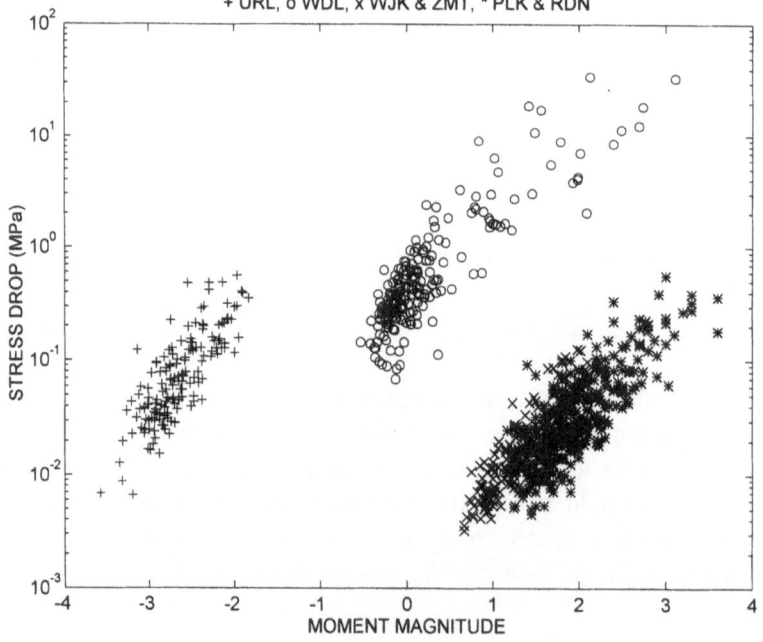

Figure 4

Stress drop versus moment magnitude for seismic events from six selected mines.

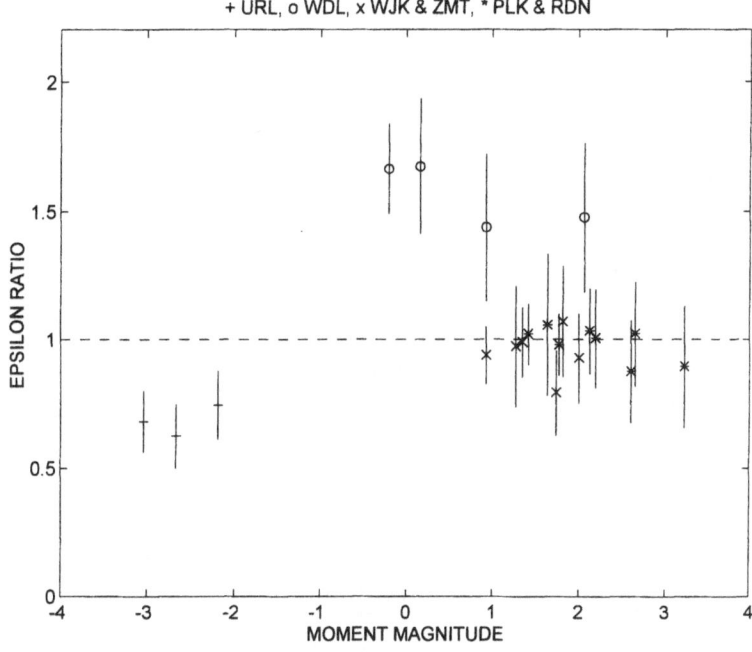

Figure 5

Mean values of the epsilon ratio versus moment magnitude for 21 sets of seismic events from six selected mines. Each set corresponds to a magnitude interval equal approximately to half the magnitude unit.

sequences, and two sequences from the PLK copper mine associated with two well-defined sections of the mine. The first PLK sequence of seismic events occurred between October 2, 1994 and May 26, 1995 in section G-21. It contains 56 well recorded events with moment magnitude ranging from 1.2 to 3.0. The second PLK sequence occurred between April 5 and August 29, 1996 in section G-51, and is composed of 47 events with moment magnitude ranging from 1.3 to 2.9. The selected seismic events from the other three Polish mines have occurred in various areas of these mines and the number of events in any particular area was too small to study their temporal trends.

A more distinct presentation of time distribution of epsilon is achieved when its values are displayed against the consecutive number of seismic events within a given sequence instead of the time of their occurrence. Furthermore, the scatter of ε values is considerable and a smoothing procedure is needed. A 10-points moving window (moved with a step corresponding to a single event) to average ε values was applied to the URL and WDL data and a 5-point window was used for the PLK data, and the mean ε values were ascribed to the interval center. The results of such a procedure for seismic events from the WDL and URL mines are shown in Figure 6 and the two sequences from the PLK mine are presented in Figure 7, where the confidence levels equal to the standard deviation of a single observation are also marked, and where the vertical bars indicate the time of occurrence of the largest events in a given sequence.

The time distributions of ε values for seismic events from all four sequences exhibit some kind of regularity and imply that the largest events in a given sequence occur when the values of ε are low and a partial stress drop mechanism is dominant. The largest events are then followed almost immediately by the high values of ε corresponding to a frictional overshoot mechanism. The events from the WDL mine display, of course, only frictional overshoot, although the time distribution of their ε values are of the same pattern (Fig. 6) as that from the other mines.

Such analyses must be repeated with more data sets to confirm or refute the described behavior of ε, nonetheless such systematic changes, if they are true, would be highly important. Unfortunately, proper data are not readily available and the problem is only touched upon here and awaits further research.

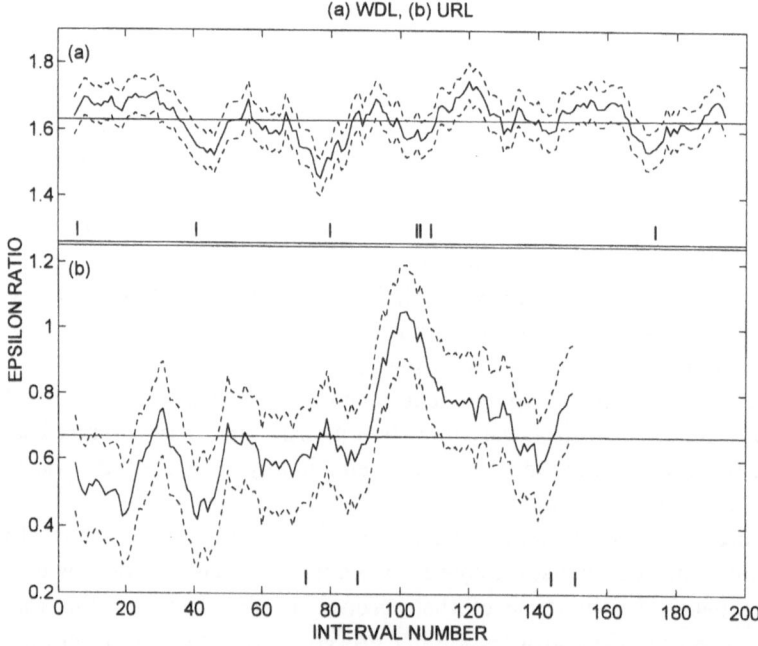

Figure 6

Mean values of the epsilon ratio from ten successive events versus the consecutive number of the interval containing the ten events (moved with a step corresponding to a single event), for seismic events from the Western Deep Levels gold mine (a) and the Underground Research Laboratory (b). The confidence levels equal to the standard deviation of a single observation are marked by dashed lines. The vertical bars indicate the time of occurrence of seismic events with magnitude greater than 2.0 at WDL and greater than -2.0 at URL.

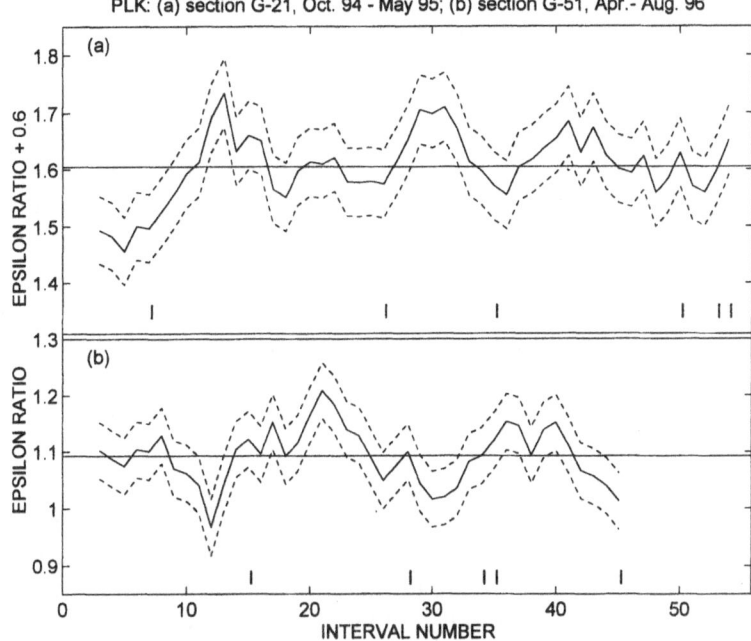

Figure 7

Mean values of the epsilon ratio from five successive events versus the consecutive number of the interval containing the five events (moved with a step corresponding to a single event), for seismic events from section G-21 (a) and G-51 (b) of the Polkowice copper mine. The confidence levels equal to the standard deviation of a single observation are marked by dashed lines. The vertical bars indicate the time of occurrence of seismic events with magnitude greater than 2.5. The ε values for the events from section G-21 are moved up by 0.6 for the sake of clarity of their presentation with the ε values form section G-51.

Conclusions

If the BRUNE (1970, 1971) stress drop is accepted as the static stress drop, then the quantity ε proposed by ZÚÑIGA (1993) can be used as an indicator of stress drop mechanism of small seismic events induced by mining. The events induced at the Underground Research Laboratory in Manitoba by shaft sinking in a granitic rock, performed by a full-face drill-and-blast technique, are characterized by low values of epsilon corresponding to a partial stress drop mechanism. A period of increased seismicity was observed after each blast and the observed mechanism could possibly be explained by highly fractured rock mass where the events were generated. In contrast, the events induced at the Western Deep Levels hard-rock gold mine in South Africa occurred at a depth of 3 km in a highly competent rock mass and they display frictional overshoot mechanisms. The stress release mechanisms of seismic events at Polish medium-depth coal and copper mines are in good agreement with the OROWAN (1960) condition such that the final stress is equal to the dynamic frictional stress, regardless of different mining techniques used and different rock mass structure present in coal and copper mines.

The static stress drop represented by BRUNE's stress drop is heavily model dependent through the source radius-corner frequency relation. For the events from the Underground Research Laboratory, the OROWAN's condition would be fulfilled if a constant in the source dimension-corner frequency relation is equal to 1.82, which is the middle value between the average value of 1.32 corresponding to S waves in the source model of MADARIAGA (1976) and that of BRUNE (1970; 1971). The constant in the MADARIAGA model, however, depends on the rupture velocity and the azimuth of observation, and if the obtained value of 1.82 could be explained in these terms, the partial stress drop mechanism there would be irrelevant. For the events from Western Deep Levels, on the other hand, the value of 3.92 of the same constant is too high for a reasonable explanation in terms of fault and observation geometry or rupture velocity, implying that the appearance of frictional overshoot mechanism there is real.

The time distributions of quantity ε for seismic events from four selected sequences display characteristic regularity. The largest events in a given sequence occur when the values of epsilon are low and a partial stress drop mechanism is dominant. These events are then followed by the high ε values corresponding to a frictional overshoot mechanism. Such analyses must be repeated with more data sets to confirm or to refute the described behavior of ε, but such changes, if they are systematic, would be highly important.

Acknowledgment

I am grateful to an anonymous reviewer whose comments helped to clarify several points in this paper.

REFERENCES

BOATWRIGHT, J. (1984) *Seismic Estimates of Stress Release*, J. Geophys. Res. *89*, 6961–6968.

BRUNE, J. N. (1970), *Tectonic Stress and the Spectra of Seismic Shear Waves from Earthquakes*, J. Geophys. Res. *75*, 4997–5009.

BRUNE, J. N. (1971), *Correction*, J. Geophys. Res. *76*, 5002.

BRUNE, J. N. *The physics of earthquake strong motion*. In *Seismic Risk and Engineering Decisions* (eds. Lomnitz, C., and Rosenblueth, E.) (Elsevier, New York 1976) pp. 141–177.

CASTRO, R. R., PACOR, F., and PETRUNGARO, C. (1997), *Determination of S-wave Energy Release of Earthquakes in the Region of Friuli, Italy*, Geophys. J. Int. *128*, 399–408.

DEICHMANN, N. (1997), *Far-field Pulse Shapes from Circular Sources with Variable Rupture Velocities*, Bull. Seismol. Soc. Am. 87, 1288–1296.

FEIGNIER, B., and YOUNG, R. P. (1992), *Moment Tensor Inversion of Induced Microseismic Events: Evidence of Non-shearing Failures in the $-4 < M < -2$ Moment Magnitude Range*, Geophys. Res. Lett. *19*, 1503–1506.

GIBOWICZ, S. J., *The mechanism of large mining tremors in Poland*. In *Rockbursts and Seismicity in Mines* (eds. Gay, N. C., and Wainwright, E. H.) (South African Institute of Mining and Metallurgy, Johannesburg 1984) pp. 17–28.

GIBOWICZ, S. J. (1995), *Scaling Relations for Seismic Events Induced by Mining*, Pure appl. geophys. *144*, 191–209.

GIBOWICZ, S. J. (1996), *Relations between Source Mechanism and the Ratio of S- over P-wave Energy for Seismic Events Induced by Mining*, Acta Montana, Ser. A. *9* (100), 7–15.

GIBOWICZ, S. J. (1997a), *Scaling Relations for Seismic Events at Polish Copper Mines*, Acta Geophys. Pol. *45*, 169–181.

GIBOWICZ, S. J. (1997b), *An Anatomy of a Seismic Sequence in a Deep Gold Mine*, Pure appl. geophys. *150*, 393–414.

GIBOWICZ, S. J., and KIJKO, A., *An Introduction to Mining Seismology* (Academic Press, San Diego 1994).

GIBOWICZ, S. J., HARJES, H.-P., and SCHÄFER, M. (1990), *Source Parameters of Seismic Events at Heinrich Robert Mine, Ruhr Basin, Federal Republic of Germany: Evidence for Nondouble-Couple Events*, Bull. Seismol. Soc. Am. *80*, 88–109.

GIBOWICZ, S. J., YOUNG, R. P., TALEBI, S., and RAWLENCE, D. J. (1991), *Source Parameters of Seismic Events at the Underground Research Laboratory in Manitoba, Canada: Scaling Relations for Events with Moment Magnitude Smaller than −2*, Bull. Seismol. Soc. Am. *81*, 1157–1182.

MADARIAGA, R. (1976), *Dynamics of an Expanding Circular Fault*, Bull. Seismol. Soc. Am. *66*, 639–666.

MCGARR, A., *Some applications of seismic source mechanism studies to assessing underground hazard*. In *Rockbursts and Seismicity in Mines* (eds. Gay, N. C., and Wainwright, E. H.) (South African Institute of Mining and Metallurgy, Johannesburg 1984) pp. 199–208.

MCGARR, A. (1991), *Observations Constraining Near-source Ground Motion Estimated from Locally Recorded Seismograms*, J. Geophys. Res. *96*, 16,495–16,508.

MCGARR, A. (1992a), *An Implosive Component in the Seismic Moment Tensor of a Mining-induced Tremor*, Geophys. Res. Lett. *19*, 1579–1582.

MCGARR, A. (1992b), *Moment Tensors of Ten Witwatersrand Mine Tremors*, Pure appl. geophys. *139*, 781–800.

MCGARR, A. (1994), *Some Comparisons between Mining-induced and Laboratory Earthquakes*, Pure appl. geophys. *142*, 467–489.

MENDECKI, A. J., *Principles of monitoring seismic rockmass response to mining*. In *Rockbursts and Seismicity in Mines* (eds. Gibowicz, S. J., and Lasocki, S.) (Balkema, Rotterdam 1997) pp. 69–80.

OROWAN, E. (1960), *Mechanism of Seismic Faulting in Rock Deformation*, Geol. Soc. Am. Mem. *79*, 323–345.

REVALOR, R., JOSIEN, J. P., BESSON, J. L., and MAGRON, A., *Seismic and seismoacoustic experiments applied to the prediction of rockbursts in French coal mines*. In *Rockbursts and Seismicity in Mines* (ed. Fairhurst, C.) (Balkema, Rotterdam 1990) pp. 301–306.

SATO, K., and FUJII, Y. (1989), *Source Mechanism of a Large-scale Gas Outburst at Sunagawa Coal Mine in Japan*, Pure appl. geophys. *129*, 325–343.

SATO, T., and HIRASAWA, T. (1973), *Body Wave Spectra from Propagating Shear Cracks*, J. Phys. Earth *21*, 415–431.

SAVAGE, J. C., and WOOD, M. D. (1971), *The Relation between Apparent Stress and Stress Drop*, Bull. Seismol. Soc. Am. *61*, 1381–1388.

SMITH, K. D., BRUNE, J. N., and PRIESTLY, K. F. (1991), *The Seismic Spectrum, Radiated Energy, and the Savage and Wood Inequality for Complex Earthquakes*, Tectonophysics *188*, 303–320.

SNOKE, J. A. (1987), *Stable Determination of (Brune) Stress Drops*, Bull Seismol. Soc. Am. *77*, 530–538.

SNOKE, J. A., Linde, A. T., and SACKS, I. S. (1983), *Apparent Stress: An Estimate of the Stress Drop*, Bull. Seismol. Soc. Am. *73*, 339–348.

WIEJACZ, P. (1992), *Calculation of Seismic Moment Tensor for Mine Tremors from the Legnica-Glogow Copper Basin*, Acta Geophys. Pol. *40*, 103–122.

WIEJACZ, P., *Moment tensors for seismic events from Upper Silesian coal mines, Poland*. In *Mechanics of Jointed and Faulted Rock* (ed. Rossmanith, H.-P.) (Balkema, Rotterdam 1995) pp. 667–672.

WYSS, M. (1970), *Stress Estimates of South American Shallow and Deep Earthquakes*, J. Geophys. Res. 75, 1529–1544.
ZÚÑIGA, F. R. (1993), *Frictional Overshoot and Partial Stress Drop. Which One?*, Bull. Seismol. Soc. Am. 83, 939–944.

(Received March 23, 1998, revised June 30, 1998, accepted July 15, 1998)

Pure appl. geophys. 153 (1998) 21–40
0033–4553/98/010021–20 $ 1.50 + 0.20/0

Pure and Applied Geophysics

Dominant Directions of Epicenter Distribution of Regional Mining-induced Seismicity Series in Upper Silesian Coal Basin in Poland

STANISLAW LASOCKI[1] and ADAM IDZIAK[2]

Abstract—The regional mining-induced seismicity of the Upper Silesian Coal Basin, Poland forms two major and two minor spatial clusters. The directional patterns of seismic series from the major clusters were studied with the use of the analysis of deflections. The seismic series is parameterized by the deflection angle of the straight line connecting epicenters of every two consecutive events, measured from NS direction. The trends of epicenter migration are characterized by modes of distribution of the deflection angle, estimated by the nonparametric kernel method. The distribution of deflection angles for the studied seismic series is not random. Altogether four trends of epicenter migration have been identified: two are connected with the subseries of events that belong to the same cluster and are related to the shape of the clusters, whereas the other two, linked to the subseries of events that alternate between the clusters, indicate that mutual positions of events in such series are not random. The results support recent hypotheses pertaining to low tectonic instability of this region.

Key words: Epicenter migration, directional patterns, induced seismicity, nonparametric estimation.

Introduction

Located in the southwestern part of Poland and partially in the Czech Republic, the Upper Silesian Coal Basin (USCB) is a region of intensive coal mining engaged in since the 18th century. At present more than 60 active collieries are distributed over an area of less than 80×60 km^2. Mining takes place at considerable depths, varying from 700 to 900 m. Productive areas are fragmented by many faults, joints and other zones of weakness, and the coal seams are frequently overlaid by thick layers of strong and rigid sandstone strata. These factors, as well as the influence of multiple remnants of older works, provide conditions for induced seismicity generation and rockbursting. Some 40 mines of the USCB work under considerable rockburst hazard.

[1] University of Mining and Metallurgy, Department of Geophysics, al. Mickiewicza 30, 30–059 Kraków, Poland.
[2] Silesian University, Faculty of Earth Sciences, ul. Bedzinska 60, 41–200 Sosnowiec, Poland.

The mining-induced seismicity in the USCB forms two general distinct classes (GIBOWICZ, 1990; GIBOWICZ and KIJKO, 1994). Most events are directly linked to mining works, occur close to active mining stopes, and their generation is controlled by the local stress field and strength conditions of the rockmass in the vicinity of the stopes. The energy of this type of event is small to intermediate and rarely exceeds 10^6J (local magnitude 2.2). Due to their location close to the areas of mining works, the strongest events from this class often have damaging effects in the stopes.

Conversely, the events that belong to the second class are more regional in nature. Their energies are higher than 10^6J and they usually occur at considerable distances from active stopes, in a certain correlation with the location of regional geological discontinuities and zones of weakness. Therefore, this type of seismicity rarely causes rockbursts. However, it often generates damaging effects on surface. The seismicity of the second class is believed to be controlled by the regional stress build-up on more than one mine scale and triggered by mining operations (JOHNSTON and EINSTEIN, 1990; GIBOWICZ and KIJKO, 1994). It may also have a triggered nature according to the classification proposed by MCGARR and SIMPSON (1997), however the knowledge of stress drops and ambient stress changes caused by mining in the area of USCB is too insufficient to verify such a conclusion.

The epicenter distribution of events of energy $E \geq 10^6$J in this area of USCB is presented in Figure 1. The epicenters form two major (A,B) and two minor (C,D)

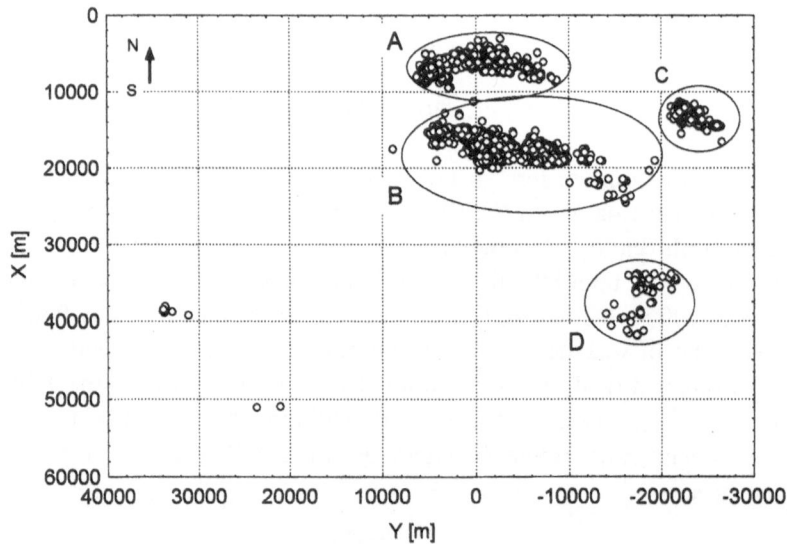

Figure 1
Epicentral distribution of four clusters of events of energy $E \geq 10^6$J in the Upper Silesian Coal Basin in Poland.

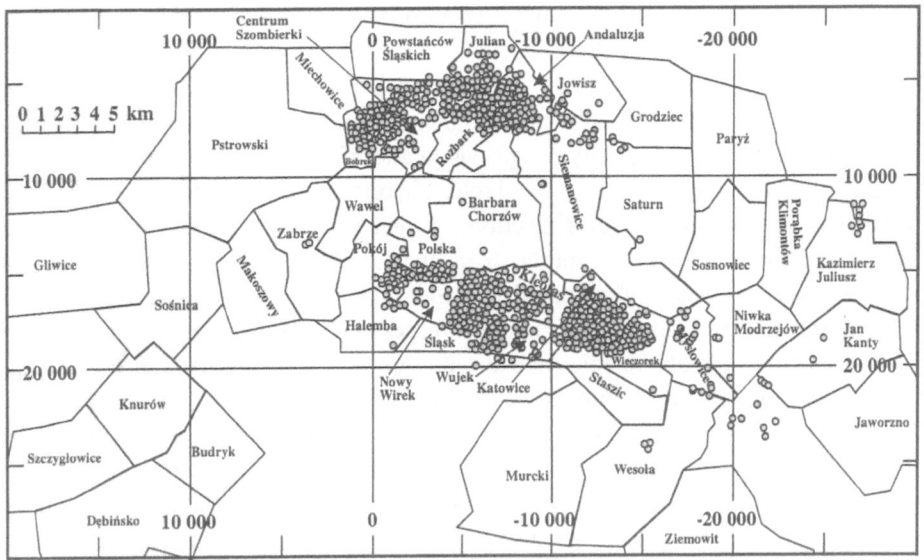

Figure 2
The location of two major clusters of seismicity of the USCB and the active mine areas.

clusters. The major clusters, whose location with respect to the active mine areas is shown in Figure 2, are some 20-km long and 4-km wide. They are neither confined to individual mines, nor cover evenly the whole area of mining works. The mines in the area of clusters as well as the mines outside this area also experience intense seismic activity of event energy below 10^6J which could be, according to the differentiation given above, classified as directly connected with mining.

The Upper Silesian Coal Basin does not experience natural seismicity. Nevertheless, recent studies of the structural geology together with the analysis of focal mechanisms of the strongest induced events in selected areas gave rise to a controversial hypothesis about possible low tectonic instability of the region (TEPER, 1990; JURA, 1996; ZUBEREK et al., 1996). In this study we investigated the seismic series that build the major clusters of seismicity in an attempt to provide additional data which could be helpful in determining a reliable seismotectonic model of the region.

In constructing such a model, seismic data are usually used to identify locations and orientations of preexisting faults and other geological features which are seismically active. Various advanced techniques were used for this purpose, including determination of fault-plane solutions of the events (e.g., GIBOWICZ 1984; McGARR, 1984; UDIAS, 1989; WIEJACZ and LUGOWSKI, 1997), the three-point method based on hypocentral distribution of events (FEHLER et al., 1987; URBANCIC et al., 1993), multiplet analysis resulting from wave-form similarity study (e.g., MORIYA et al., 1994, 1996; LESNIAK and NIITSUMA, 1996; SPOTTISWOODE and

MILEV, 1998) and others. None of the mentioned approaches could be applied to analyze the seismicity from the USCB because of the lack of required input data. The available seismic data files for the studied period from 1977 to 1994 contain only the time of event occurrences, epicenteral coordinates and energy of events. The hypocentral depth is mostly not provided and, if present, highly unreliable, and the complete wave forms are not available. Therefore, we applied another method, which we call the analysis of deflections, which could also work on the limited information provided in the catalogs from the USCB.

Our earlier studies of the spatial distribution of seismic series carried out for local mining-induced seismicity in selected stopes of Wujek Coal Mine, showed that nonrandom features of the epicenter distribution of seismic events can be assessed through the analysis of the directions of epicenter migration (LASOCKI et al., 1997). If only coordinates of epicenters are known, the seismic series is parameterized by the deflection angle of the straight line connecting every two consecutive events, measured from the NS direction. The method is based on nonparametric kernel estimation of the probability density function (pdf) of the deflection: modes of the pdf identify dominant directions of epicenter migration. By definition, the series of deflections is a time series, and the method has been basically designed to analyze time variation of the directions of source migration. However, our studies on complete series from Wujek Mine demonstrated that it can also be used to assess global directional structure of the data. The identified trends of epicenter migration could be correlated either with mining or local geological and tectonic features, providing better insight into factors responsible for seismicity generation.

Data

The studied data comprised all seismic events of energy larger than 10^6J which occurred in the USCB from 1977 to 1994, and were classified as belonging to the major clusters of regional seismicity. Altogether 3500 events were included in the data set, out of which 1680 belong to the northern and 1820 to the southern cluster, respectively. The data set was compiled from the seismic catalog of events recorded by the regional seismic network, operated by Central Mining Institute, which covers the entire area of the USCB and from the catalogs of events recorded by local mine networks, which monitor the areas of the most seismically active mines. Epicenter locations provided in these catalogs were obtained from *P*-wave arrivals on multiple sensors. Due to the difference in average distances between seismic stations in the regional and mining networks, the locations differ in accuracy, however, statistically the location error does not exceed 150 m.

The linear relation between the logarithm of energy and logarithm of the number of events in energy classes for the studied data, shown in Figure 3a, proves the completeness of the data in the entire energy range. There are no significant

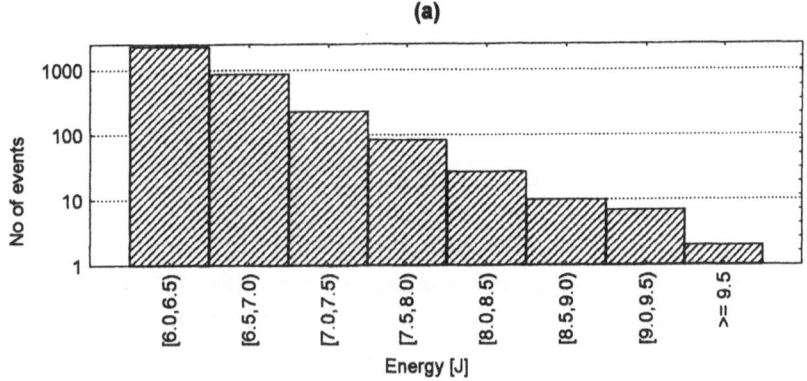

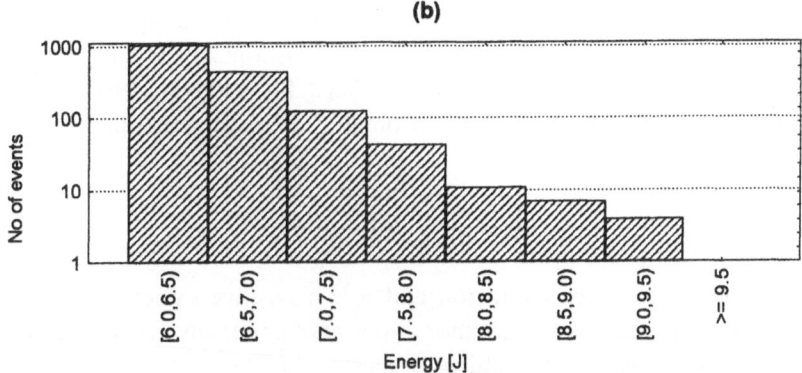

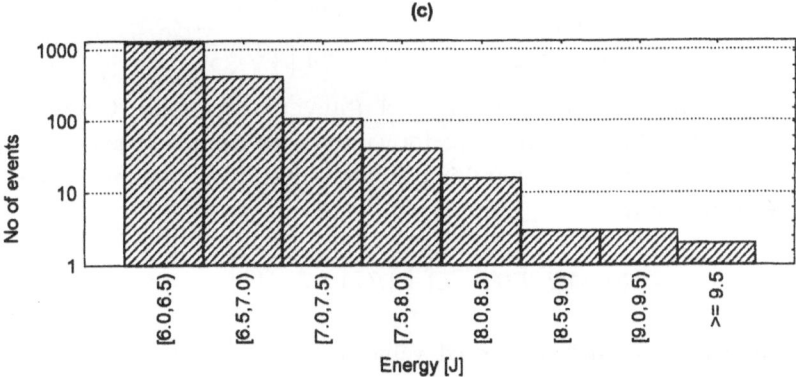

Figure 3
Histogram of energy, on a log-log scale, for the studied data: (a) the complete data set; (b) the events
from the northern cluster; (c) the events from the southern cluster.

differences between the energy distribution of events from clusters A and B, given in Figures 3b and 3c, respectively, except in the least populated high energy classes.

Method

As earlier presented, the mutual spatial orientation of every two consecutive seismic events is characterized, in our study, by the deflection angle of a straight line connecting their epicenters, measured from the NS direction. The sense of the vector connecting the two epicenters is not assessed, thus the deflection varies in the range of $[-90°, 90°]$. Given a seismic series of n elements, the deflections form a new series of $(n-1)$ directions of epicenter migration. We assume further that deflection angles are independent, identically distributed, random variables. Thus complete information concerning this parameter is stored in its probability density function.

The pdf of deflection is estimated by means of the nonparametric kernel estimation method; the technique introduced by ROSENBLATT (1956) and PARZEN (1962) and significantly developed recently (e.g., SILVERMAN, 1986). Given the sample data x_i, $i = 1, \ldots, n$, the kernel estimate $\hat{f}(x)$ of an actual pdf $f(x)$ is given by

$$\hat{f}(x) = \frac{1}{nh} \sum_{i=1}^{n} K\left(\frac{x - x_i}{h}\right), \tag{1}$$

where $K(\bullet)$ is a given kernel function and h is a positive smoothing factor. The choice of the kernel function is of minor importance and many symmetric unimodal distribution functions ensure similar efficiency of the method. We used the Epanechnikov, parabolic kernel function of the form

$$K_e(t) = \begin{cases} \frac{3}{4\sqrt{5}}\left(1 - \frac{t^2}{5}\right) & \text{for } |t| \leq \sqrt{5}, \\ 0 & \text{for } |t| > \sqrt{5} \end{cases} \tag{2}$$

whose asymptotic efficiency is equal to 1 (SILVERMAN, 1986). Conversely, the choice of h is crucial for the proper performance of the method. The most common approach to estimating the smoothing factor is based on minimizing the mean integrated square error

$$MISE(\hat{f}) = E\left[\int \{\hat{f}(x) - f(x)\}^2\right], \tag{3}$$

where $E[\bullet]$ stands for the anticipated value (e.g., ROSENBLATT, 1956; RUDEMO, 1982; HALL, 1983; BOWMAN, 1984; STONE, 1984). We applied an approximate method of minimizing of $MISE$ for the parabolic kernel, developed by ADAMOWSKI and FELUCH (1987). Under some simplifying assumptions, the smoothing factor for the parabolic kernel can be approximated by the following expression

$$h \approx \frac{\sum_{i=1}^{n} (2i - n - 1)x_i'}{n(n - 10/3)\sqrt{5}}, \qquad (4)$$

where $\{x_i'\}$ are ordered sample values. Being aware of possible difficulties with the reference written in Polish, we provide a description of this method in the Appendix. The Monte Carlo simulation studies evidenced that the method precisely locates modes of pdf, is fast for large samples and quite effective for small ones. The pdf was satisfactorily reproduced, even from samples of 40 elements. The drawback of the method is that it oversmoothes the pdf so that the modal value can be reduced and the peak width can be increased, and some secondary features of the distribution can be lost. On the other hand, these properties of the method help keep our deflection mode identification on the safe side in the sense that the chance to identify and interpret artifacts becomes quite low.

The deflection angles naturally lie within the interval $[-90°, 90°]$, which can be regarded as a circle with a circumference of 180° so that the deflection angle obeys the identity

$$\alpha \equiv \begin{cases} \alpha + 180° & \text{for } \alpha \in [-180°, -90°] \\ \alpha - 180° & \text{for } \alpha \in [90°, 180°] \end{cases}. \qquad (5)$$

The nonparametric density estimation, was originally designed for the entire real line. In order to eliminate boundary effects resulting from the finite support of the kernel estimator, we estimated the density function four times for every data sample, each time shifting the deflections by 30°. The final form of the pdf estimate was reached by joining end to end the central one quarters of every four estimates.

Securing the pdf estimate we are mostly interested in its modes and bumps since they are 'descriptive features likely to indicate mixing of components' (COX, 1966). A mode in a density is a local maximum, while a bump is an interval [a,b] such that the pdf is concave over [a,b] but not over any larger interval (SILVERMAN, 1986). In case of the deflection angle a mode means that, given the event, the probability of location of the next event close to the line passing through the epicenter of the first event, and deflected from the N-S direction of the value of mode, is greater than the probability of any other location. The distribution around the mode is characterized by the value of mode, the modal value i.e., the value of pdf for the mode, the width of peak, its skewness and other parameters, out of which the value of mode is our primary target. Other parameters, in particular the probabilities for a narrow range around the mode, have no straightforward meaning for the following reasons. First, such probabilities depend on the modal value, the width of the peak, as well as the number and separation of modes. Consider the unimodal distribution $f(x) = 0.6 \times U(x; -90, 90) + 0.4 \times N(x; 0, 20)$, where $U(x)$ stands for uniform distribution over the range $[-90, 90]$, and $N(x)$ denotes normal distribution with standard deviation $\sigma = 20$, which is quite a narrow peak with which 40% of the data are associated. For such a distribution the associated probability for the range of length 10 around the mode is 0.11 which means less than 6% of the raise of probability with respect to that for the uniform distribution. For the multimodal distribution with similar, well separated modes, this increase will obviously be reduced by the factor equal to the number of modes. Second, we deal with projections of source locations on the horizontal plane. In the simplest case of all sources located on one plane, the width of a mode of deflection will depend upon the dip angle of the plane and will vary from zero for the vertical plane to infinity for the horizontal plane. In the latter case we will have uniform distribution of deflection. However, in every instance, except the horizontal case, the value of mode will determine the strike direction regardless of the width of peak. Thus, the main concern is an objective identification of modality of the pdf estimate. A description of techniques employed to study the modality of nonparametric density estimates is far beyond the scope of this work. It can be found in the referenced book of SILVERMAN (1986) and in many recent papers (e.g., FISHER et al., 1994; JONES et al., 1996; CHAUDHURI and MARRON, 1997). We can only comment that, according to our experience from various Monte Carlo tests, modes of deflection, whenever distinct, pass verifying procedure positively. A detailed description of the analysis of deflections, together with studies of efficiency and sensitivity, will be given in a subsequent paper.

Identification of modes of deflection somewhat differs from identifying locations of seismically active geological features. The mode of deflection assigns the direction rather than the location of a line around which the epicenters could cluster. Therefore, the mode can, but does not have to, correlate with an apparent linear feature of epicentral distribution. An example of this possible difference between the epicentral distribution and its directional structure achieved from the analysis of deflections, taken from studies of local seismicity in Wujek Coal Mine (LASOCKI and KUSTOWSKI, 1998) is shown in Figure 4. The epicenters of a series of 50 events form a random cloud with no distinct alignments (Fig. 4a). Conversely, the distribution of deflection has a distinct and narrow mode at some 0° (Fig. 4b).

Moreover, the deflection angle is defined on a pair of consecutive events, not on all possible pairs of epicenters, and the mode of deflection accounts for the dominant direction of epicenter migration which does not have to coincide with the direction of the long axis of the epicenter distribution. Imagine a series of $m \times n$ points in which consecutive subseries of n elements are distributed ran-

(a)

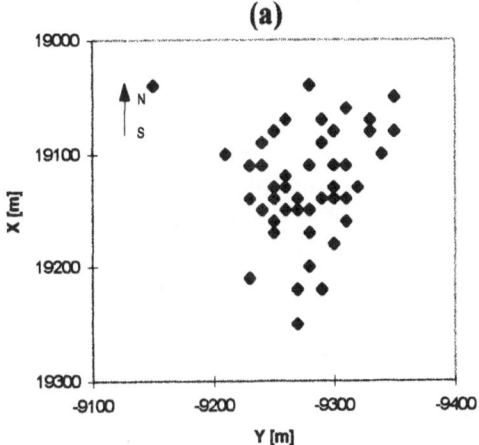

(b)

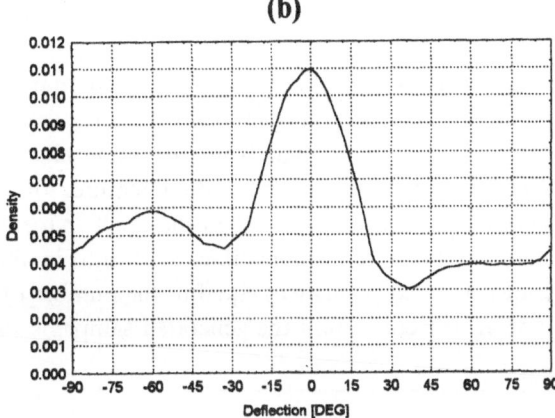

Figure 4
(a) The epicentral distribution, and (b) the probability density function of deflection, for a local series
of 50 events recorded in Wujek Coal Mine (from LASOCKI and KUSTOWSKI, 1998).

domly on parallel lines. Let the separation between the lines be greater than the
range of variability of point position on each line. A structure like this is apparently
elongated in the direction perpendicular to the lines while the analysis of deflections
will identify a very strong dominant direction parallel to the lines. For the given
reasons we interpret the modes of deflection as a directional feature of the
underlying event generating process rather than in terms of the geometry of
seismically active structures, although, in many cases, this may be the same.
Basically, however, the appearance of deflection mode indicates the existence of a
certain directional tendency in the generation of seismic events.

Data Analysis

The analysis of deflections was performed on both the complete seismic series from the studied area and period, and on its three subseries linked to a particular origin of the data. The first two subseries comprised pairs or longer time ordered groups of events in which all members belong to the same cluster: the northern (A) and the southern (B), respectively (Fig. 1). Further on, we call them in-cluster series. The third subseries was composed of all groups of consecutive events in which every member is located in a different cluster than its predecessor. It is called between-clusters series. The northern in-cluster, southern in-cluster and between-clusters series possessed 957, 1097 and 1445 elements, respectively.

Figure 5 shows the estimate of pdf of deflection for the complete series. Two distinct peaks are visible. The first mode is at some $-83°$ which broadly matches the averaged direction of long axes of both clusters. The second mode at some $-15°$ points in a direction nearly transverse to the long axis of the northern cluster. No simple geometrical justification for the presence of the second mode can be found.

Previous qualitative studies of the same data suggested that there may be a certain nonrandom link between the clusters so that an event is more likely followed by an event from the other than from the same cluster (IDZIAK and LASOCKI, 1997). In order to investigate nonrandom features of both epicentral distribution and time ordering of events we picked the exact shape of both clusters and generated a random sample of the positions of points in the areas of the clusters. The number of generated points was equal to the number of seismic events in the clusters. The pdf of deflection from the generated sample is shown in Figure

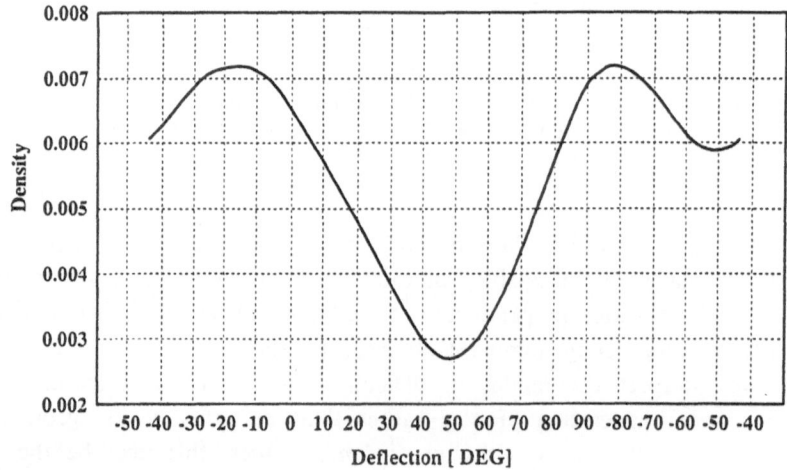

Figure 5
The probability density ·function of deflection for the complete seismic series.

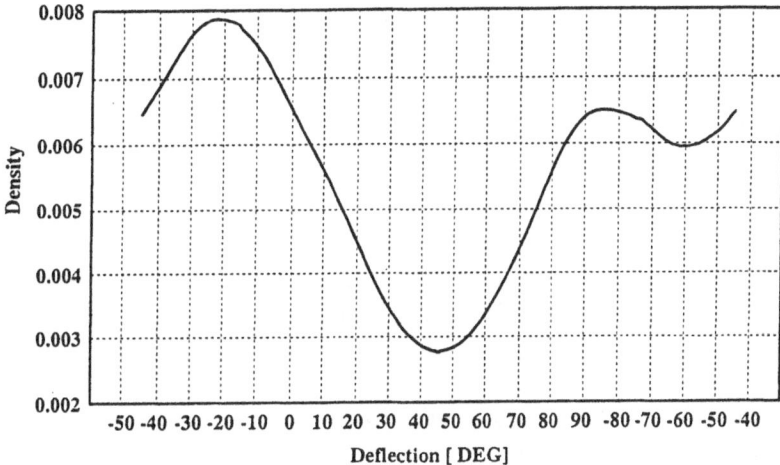

Figure 6
The probability density function of deflection for the Monte Carlo-generated locations over the areas of clusters.

6. The density is again bimodal and its modes are the same as for the pdf of deflection of the actual seismic series. However, the magnitudes of peaks are now different as the modal value connected with the transverse direction is greater.

The estimates of pdf of deflection for the in-cluster series are shown in Figure 7. The series associated with individual clusters have only one dominant direction. It amounts to some $+85°$ for the northern cluster data (Fig. 7a) and some $-75°$ for the southern cluster data (Fig. 7b). The trend of deflection at $-83°$, which was found for the complete series, is probably an average of these dominant directions. The distribution of deflection for the series of events that alternate between the clusters, presented in Figure 8, has a more complex character. The density function has a strong and wide peak in the range of $-45°$ to $+20°$, composed of two individual peaks. The modes can be identified at some $-23°$ and $+2°$, respectively. The distribution is not symmetrical and the mode at $-23°$ is definitely stronger than the other one, and thus we can suspect that the trend of $-15°$ for the complete series is a somewhat weighted average of the modes found for the between-clusters series. To test whether this shape of distribution results only from the shapes of the clusters or is due to some other feature of event generating process, we extracted from our above described random series a random between-clusters subseries and repeated the analysis of deflections for these data. The estimate of pdf of deflection is shown in Figure 9. The distribution is apparently unimodal with the mode at $-17°$. We can conclude from this experiment that the two modes identified for the real between-clusters subseries represent nonrandom features of generation of alternating between-clusters succession of events.

The distribution of all identified dominant directions is shown in Figure 10. Two modes of the in-cluster series match the directions of elongation of clusters, whereas the other modes cannot be associated with the geometry of clusters.

Discussion

A mode of the probability density function for a random variable indicates that the generating mechanism for this variable is not random. The multimodality of the pdf usually results from mixing up outcomes of more than one generating mechanism. We have identified two modes of deflection pdf for the considered complete seismic series. One of these modes is the average direction identified in the subseries associated with individual clusters, matching the directions of long axes of the

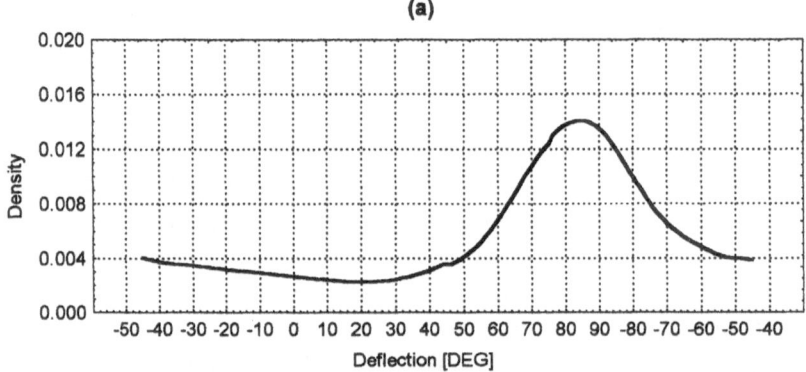

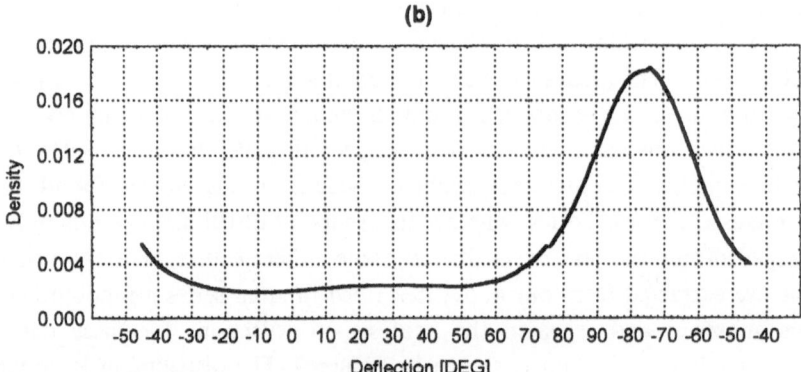

Figure 7
The probability density functions of deflection for: (a) the in-cluster series of the northern cluster; (b) the in-cluster series of the southern cluster.

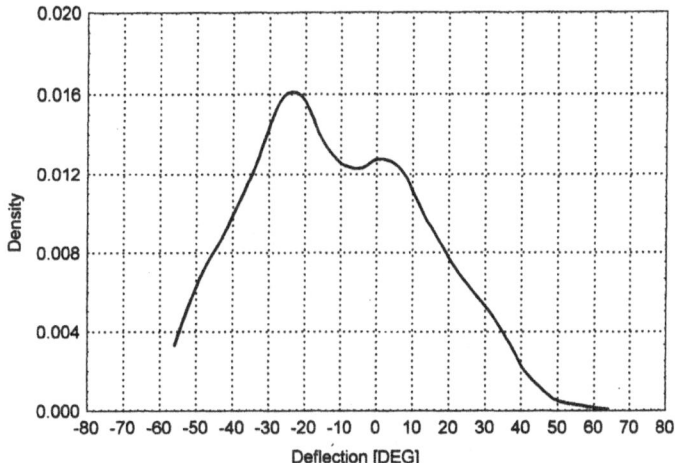

Figure 8
The probability density functions of deflection for the series of events alternating between clusters.

respective clusters. Hence, the analysis of deflection confirms the trend that can be concluded from the geometry of the clusters, outlining that in the in-cluster series the distribution of seismic sources is not random and the epicenters tend to align in the direction of the long cluster axis. The second trend identified in the complete series is connected with the series of events alternating between clusters. If two consecutive events are to be located in different clusters, the position of the second event, with respect to the first one, is governed by the bimodal density function of deflection.

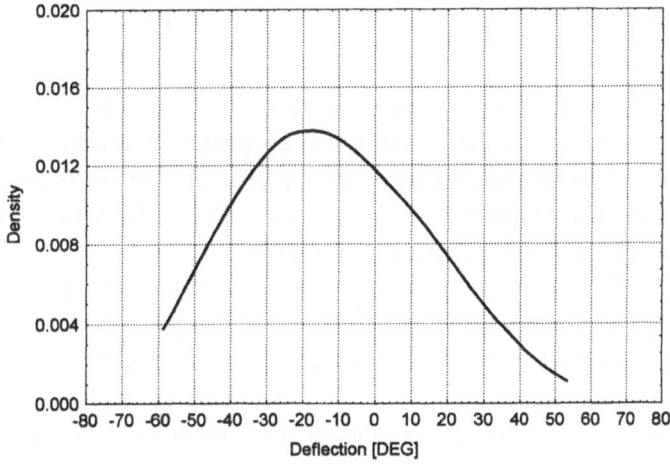

Figure 9
The probability density function of deflection for the series of events alternating between clusters, extracted from the Monte-Carlo generated locations over the areas of clusters.

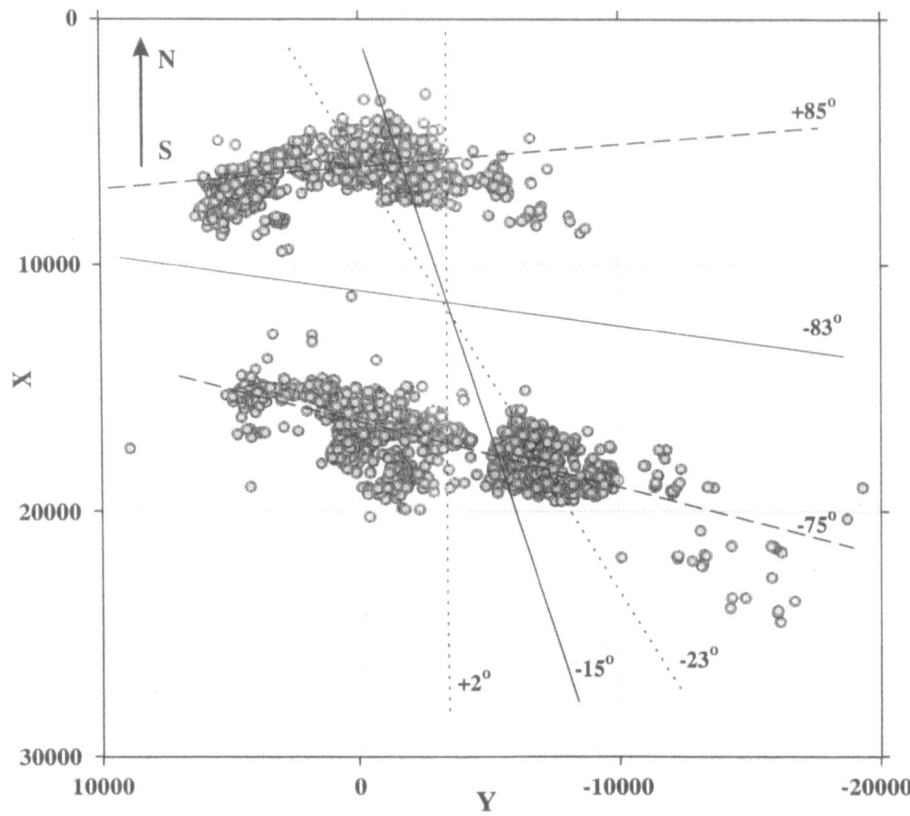

Figure 10
The identified dominant directions of source migration. Heavy line—the trends for the complete series.
Dashed line—the trends for the in-cluster series. Dotted line—the trends from the series of events
alternating between clusters.

The Monte Carlo simulation experiment suggested that both dominant trends of
the complete series are linked to the shape of clusters, whereas the position of
epicenters in the cluster areas seems to be random. The observed decrease of the
transverse mode contribution to the deflection pdf for the actual seismic data,
compared to that for the Monte Carlo sample, argues against the hypothesis
regarding a preference of the alternating between clusters of seismicity over the
in-cluster seismicity. However, the results of the analysis of deflections for the
generated between-clusters series, compared with those for the actual series, have
proved that mutual positions of consecutive epicenters in the alternating between
clusters successions of events are not random. Two dominant directions of epicenter
migration in this series have been found.

The recent studies of the geology and focal mechanisms of the strongest seismic
events of the Upper Silesian Coal Basin indicated that large discontinuities of the

USCB are related to a block structure of the deep crustal basement which has exhibited tectonic activity since Tertiary (TEPER, 1990; ZUBEREK *et al.*, 1996). At present, the existing shear stresses may cause the instability of shallow fault zones and influence the generation of strong tremors in these areas. (IDZIAK *et al.*, 1997). Moreover, it is hypothesized that the block located in the central part of the studied region, whose southern boundary is determined by a deep-rooted dislocation of the basement under the Klodnicki fault, the principal regional dislocation, has undergone both oblique E-W, ESE-WNW horizontal movements as well as an uplift (JURA, 1996). The hypothetical neogene pattern of geodynamics of the studied area and the location of the seismicity clusters are shown in Figure 11. The location of the southern cluster strictly correlates with the position of the Klodnicki fault, and the dominant direction of epicenter migration in this cluster (Fig. 10) matches the

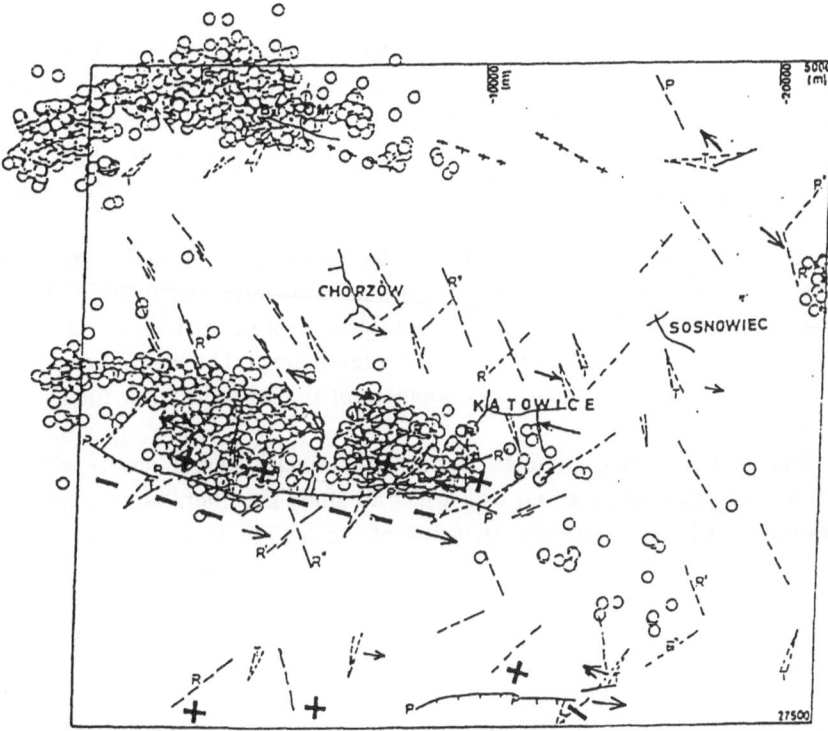

Figure 11

Neogene pattern of geodynamics of the studied area (from Jura, 1996). → tendency of sinistral strike-slip movements; + + halfhorst or elevation zones; ‒ ‒ halfgraben; Fault related: T—extension fractures, R—complementary shears, P—synthetic shears subparallel to the fault; Circles—epicenters of studied seismic events.

strike direction of the fault. The northern cluster of seismicity with its dominant direction of epicenter migration can delineate the northern boundary of this block. Apart from the seismic data there is no other clear evidence of this feature. Further developments of the present or a similar seismotectonic model of this area are also needed to explain the presence of the dominant directions of migrations for the between-clusters series.

Conclusions

1. Spatial distribution of mining triggered seismicity of the Upper Silesian Coal Basin is not random. The analysis of deflections of the seismic data from two major clusters of seismicity demonstrates that there are at least four dominant directions of epicenter migration. Two of these trends matching the direction of elongation of clusters, respectively, are connected with the in-cluster time series of events, and the other two are linked to the alternating between clusters series of events and are transverse to the cluster long axes.

2. The dominant directions of the in-clusters series are related only to the shape of the clusters. The epicenter locations within the clusters seem to be random. Conversely, when two consecutive events are to be located in different clusters, the position of the epicenter of the second event, with respect to the first one, is governed by the bimodal density function of deflection, whose modes do not result from the cluster shape.

3. The dominant directions of epicenter distribution identified from the in-cluster seismic series support the recent hypotheses addressing the internal structure and low tectonic instability of the deep crustal basement of the USCB and its effect on the stress field and stability of the shallow fault zones. The identified transverse trends should be used for further development of the seismotectonic model of the area.

4. The analysis of deflections based on nonparametric kernel density estimation has shown its potential as a tool with which to study the nonrandom character of the spatial distribution and serial structure of the seismic data.

Acknowledgments

The authors wish to thank Stanislaw Weglarczyk, of Technical University of Cracow for his assistance in data analysis. This work was sponsored by the Polish State Committee for Scientific Research under contract No. 9 T12B 00609 during the period of 1995–1998. We would also like to thank two anonymous reviewers for their comments which enhanced this manuscript.

Appendix: Approximate Estimation of Smoothing Factor for the Parabolic Kernel Function

Given a sample of n elements of the random variable X, $x_1, \ldots, x_n$, the kernel estimator of the probability density function of X is given by (1). The ideal value of h, from the point of view of minimizing the asymptotic approximation of the mean integrated square error (3) is equal to:

$$h_{\mathrm{opt}} = \alpha(K)\beta(f)n^{-1/5}, \tag{6}$$

and

$$\alpha(K) = \left\{ \int_{-\infty}^{\infty} [K(t)]^2 \, dt \right\}^{1/5} \left\{ \int_{-\infty}^{\infty} K(t)t^2 \, dt \right\}^{-2/5} \tag{7}$$

depends only upon the kernel, and

$$\beta(f) = \left\{ \int_{-\infty}^{\infty} [f''(x)]^2 \, dx \right\}^{-1/5} \tag{8}$$

depends only upon the unknown density (PARZEN, 1962).

The following approximate manner of estimating h_{opt} for the parabolic kernel function (2) was formulated by ADAMOWSKI and FELUCH (1987). From (2) and (7) we obtain for the parabolic kernel

$$\alpha(K_e) = \left(\frac{3}{5\sqrt{5}} \right)^{1/5}. \tag{9}$$

With regard to $\beta(f)$, the authors propose to replace the second derivative of the unknown density with the second derivative of its kernel estimator, which, for the considered kernel, is given by

$$\hat{f}''(x) = \frac{3}{10\sqrt{5}nh^3} \sum_{i=1}^{n} R\left(\frac{x - x_i}{h\sqrt{5}} \right), \tag{10}$$

where $R(t) = \begin{cases} 1 \text{ for } |t| < 1 \\ 0 \text{ for } |t| \geq 1 \end{cases}$.

Thus

$$\int_{-\infty}^{\infty} [\hat{f}''(x)]^2 \, dx = \frac{9}{500n^2h^6} \int_{-\infty}^{\infty} \left\{ \sum_{i=1}^{n} R\left(\frac{x - x_i}{h\sqrt{5}} \right) \right\} \left\{ \sum_{j=1}^{n} R\left(\frac{x - x_j}{h\sqrt{5}} \right) \right\} dx. \tag{11}$$

Further on it is assumed that the sample is ordered so that $x_1 \leq x_2 \leq \ldots \leq x_n$. In this case

$$\left\{ \sum_{i=1}^{n} R\left(\frac{x-x_i}{h\sqrt{5}}\right) \right\} \left\{ \sum_{j=1}^{n} R\left(\frac{x-x_j}{h\sqrt{5}}\right) \right\}$$

$$= \sum_{i=1}^{n} \left[R\left(\frac{x-x_i}{h\sqrt{5}}\right) \right]^2 + 2 \sum_{i=2}^{n} \sum_{j<i} R\left(\frac{x-x_i}{h\sqrt{5}}\right) R\left(\frac{x-x_j}{h\sqrt{5}}\right). \qquad (12)$$

It is easy to show that if $j < i$

$$R\left(\frac{x-x_i}{h\sqrt{5}}\right) R\left(\frac{x-x_j}{h\sqrt{5}}\right) = \begin{cases} 1 & \text{for } x_i - h\sqrt{5} < x < x_j + h\sqrt{5} \\ 0 & \text{elsewhere} \end{cases} . \qquad (13)$$

Thus from (11), (12) and (13)

$$\int_{-\infty}^{\infty} [\hat{f}''(x)]^2 \, dx = \frac{9}{500 n^2 h^6} \left\{ \sum_{i=1}^{n} 2\sqrt{5} h + 2 \sum_{i=2}^{n} \sum_{j=1}^{i-1} \max[0, 2\sqrt{5}h - (x_i - x_j)] \right\}$$

$$\geq \frac{9}{250 n^2 h^6} \left\{ n h\sqrt{5} + \sum_{i=2}^{n} \sum_{j=1}^{i-1} [2\sqrt{5}h - (x_i - x_j)] \right\}$$

$$= \frac{9}{250 n^2 h^6} \left\{ n^2 h\sqrt{5} - \sum_{i=1}^{n} (2i - n - 1) x_i \right\}. \qquad (14)$$

Finally, we derive from (6), (9) and (14) for the approximation of h_{opt}

$$\tilde{h}_{opt} \leq \frac{\sum_{i=1}^{n} (2i - n - 1) x_i}{n(n - 10/3) \sqrt{5}}. \qquad (15)$$

In practical applications the equality in (15) is accepted, which results in a certain amount of oversmoothing of the density estimates.

REFERENCES

ADAMOWSKI, K., and FELUCH, W. (1987), *The Comparison of Parametric and Nonparametric Methods of Flood Frequency Estimation*, Wiad. IMGW 31 (2-3), 67–78 (in Polish).

BOWMAN A. W. (1984), *An Alternative Method of Cross-validation for the Smoothing of Density Estimates*, Biometrica *71*, 353–360.

CHAUDHURI, P., and MARRON, J. S. (1997), *SiZer for Exploration of Structures in Curves*, Pers. Comm.

COX, D. R. (1966), *Notes on the Analysis of Mixed Frequency Distributions*, Brit, J. Math. Statist. Psychol. *19*, 39–47.

FEHLER, M., HOUSE, L., and KAIEDA, H. (1987), *Determining Planes along which Earthquakes Occur: Method and Application to Earthquakes Accompanying Hydraulic Fracturing*, J. Geophys. Res. *92*, 9407–9414.

FISHER, N. I., MAMMEN, E., and MARRON, J. S. (1994), *Testing for Multimodality*, Comput. Statist. and Data Anal. *18*, 499–512.

GIBOWICZ, S. J., *The mechanism of large mining tremors in Poland*. In *Rockbursts and Seismicity in Mines* (eds. Gay, N. C., and Wainwright, E. H.) (S. Afr. Inst. Min. Metall., Johannesburg 1984) pp. 17–28.

GIBOWICZ, S. J., *Keynote lecture: The mechanism of seismic events induced by mining—a review*. In *Rockbursts and Seismicity in Mines* (ed. Fairhurst, C.) (Balkema, Rotterdam 1990) pp. 3–27.

GIBOWICZ, S. J., and KIJKO, A., *An Introduction to Mining Seismology* (Academic Press, San Diego 1994).

HALL, P. (1983), *Large Sample Optimality of Least-squares Cross-validation in Density Estimation*, Ann. Statist. *11*, 1156–1174.

IDZIAK, A., and LASOCKI, S., *Studies of serial structure of induced seismicity from Upper Silesian Coal Basin*. In *Results from Recent Study in Seismology and Engineering Geophysics—Proc. Regional Conference with International Participation Ostrava 8–9 April, 1997* (ed. Kalab, Z.) (Academy of Sciences of Czech Republic, Institute of Geonics, Prague 1997) pp. 151–158 (in Polish).

IDZIAK, A., TEPER, L., Zuberek, W. M., SAGAN, G., and DUBIEL, R., *Mine tremor mechanisms used to estimate the stress field near the deep-rooted fault in the Upper Silesian Coal Basin, Poland*. In *Rockbursts and Seismicity in Mines* (eds. Gibowicz, S. J., and Lasocki, S.) (Balkema, Rotterdam 1997) pp. 31–37.

JOHNSTON, J. C., and EINSTEIN, H. H., *A survey of mining associated rockbursts*. In *Rockbursts and Seismicity in Mines* (ed. Fairhurst, C.) (Balkema, Rotterdam 1990) pp. 121–127.

JONES, M. C., MARRON, J. S., and SHEATHER, S. J. (1996), *A Brief Survey of Bandwidth Selection for Density Estimation*, J. Am. Statist. Assoc. *91*, 401–407.

JURA, D., *Young Alpine stress field in the Bytom-Katowice plateau northern part of the Upper Silesian Coal Basin*. In *Tectonophysics of Mining Areas* (ed. Idziak, A.) (Wyd. Uniw. Slaskiego, Katowice 1996) pp. 29–40.

LASOCKI, S., and KUSTOWSKI, B., *Seismic hazard and dominant directions of source migration during longwall mining: Examples from Wujek Coal Mine*. In *Proc, Conf. Rockbursts '98—Safe Mining* (Central Mining Institute, Katowice 1998) in print (in Polish).

LASOCKI, S., WEGLARCZYK, S., and GIBOWICZ, S. J., *A new method to estimate directional character of mining induced seismicity: Application to the data from Wujek Coal Mine, Poland*. In *Rockbursts and Seismicity in Mines* (eds. Gibowicz, S. J., and Lasocki, S.) (Balkema, Rotterdam 1997) pp. 207–211.

LESNIAK, A., and NIITSUMA, H., *Clustering similar AE events using the filtered wave-form envelope*. In *Progress in Acoustic Emission VIII* (eds. Kishi, T., Mori, Y., Higo, H., and Enoki, M.) (The Japanese Society for NDI, Nara 1996) pp. 133–140.

MCGARR, A., *Some applications of seismic source mechanism studies to assessing underground hazard*. In *Rockbursts and Seismicity in Mines* (eds. Gay, N. C., and Wainwright, E. H.) (S. Afr. Inst. Min. Metall., Johannesburg 1984) pp. 199–208.

MCGARR, A., and SIMPSON, D,. *Keynote lecture: A broad look at induced and triggered seismicity*. In *Rockbursts and Seismicity in Mines* (eds. Gibowicz, S. J., and Lasocki, S.) (Balkema, Rotterdam 1997) pp. 385–396.

MORIYA, H., NAGANO, K., and NIITSUMA, H. (1994), *Precise Source Location of AE Doublets by Spectral Matrix Analysis of Triaxial Hodogram*, Geophysics *59*, 36–45.

MORIYA, H., RUTLEDGE, J. T., KAEIDA, H., and NIITSUMA, H., *Subsurface stress field determination using multiplets in downhole three-component microseismic measurement*. In *Proc. 2nd North American Rock Mechanics Symposium: NARMS'96* (eds. Aubertin, M., Hassani, F., and Mitri, H.) (Balkema, Rotterdam 1996) pp. 853–858.

PARZEN, E. (1962), *On Estimation of Probability Density Function and Mode*, Ann. Math. Statist. *33*, 1065–1076.

RUDEMO, M. (1982), *Empirical Choice of Histograms and Kernel Density Estimators*, Scand. J. Statist. *9*, 65–78.

ROSENBLATT, M. (1956), *Remarks on Nonparametric Estimates of a Density Function*, Ann. Math. Statist. *27*, 832–835.

SILVERMAN, B. W., *Density Estimation for Statistics and Data Analysis* (Chapman and Hall, London 1986).

SPOTTISWOODE, S. M., and MILEV, A. M. (1998), *The Use of Wave-form Similarity to Define Planes of Mining-induced Seismic Events*, Tectonophysics *289*, 51–60.

STONE, C. J. (1984), *An Asymptotically Optimal Window Selection Rule for Kernel Density Estimates*, Ann. Statist. *12*, 1285–1297.

TEPER, L. (1990), *Meso- and Macrotectonic Indicators of Strike-slip Movements in Crystalline Basement of NE Part of the USCB*, Papers Com. Pol. Acad. Sci. *14*, 40–41.

UDIAS, A., *Development of fault-plane studies for the mechanism of earthquakes*. In *Observatory Seismology* (ed. Litehiser, J. J.) (University of California Press, Berkeley 1989) pp. 243–356.

URBANCIC, T. I., TRIFU, C.-I., and YOUNG, R. P. (1993), *Microseismicity Derived Fault-planes and their Relationship to Focal Mechanism, Stress Inversion, and Geologic Data*, Geophys. Res. Lett. *20*, 2475–2478.

WIEJACZ, P., and LUGOWSKI, A., *Effects of geological and mining structures upon mechanism of seismic events at Wujek coal mine, Katowice, Poland*. In *Rockburst and Seismicity in Mines* (eds. Gibowicz, S. J., and Lasocki, S.) (Balkema, Rotterdam 1997) pp. 27–30.

ZUBEREK, W. M., TEPER, L., IDZIAK, A., and SAGAN, G., *Tectonophysical approach to the description of mining-induced seismicity in the Upper Silesia*. In *Tectonophysics of Mining Areas* (ed. Idziak, A.) (Wyd. Uniw. Slaskiego, Katowice 1996) pp. 79–98.

(Received June 30, 1998, accepted August 21, 1998)

Pure appl. geophys. 153 (1998) 41–65
0033–4553/98/010041–25 $ 1.50 + 0.20/0

┃Pure and Applied Geophysics

Use of Microseismic Source Parameters for Rockburst Hazard Assessment

JANE M. ALCOTT,[1] PETER K. KAISER[1] and BRAD P. SIMSER[2]

Abstract—Since 1994 Noranda's Brunswick #12 Mine has complemented their MP250/Queen's Full Waveform seismic systems with an ISS (Integrated Seismic System). Time histories of ISS source parameter information form a component of the daily ground control decision-making. This paper discusses a methodology for microseismic hazard assessment, which filters ISS data using energy, apparent stress and seismic moment criteria to identify those events that are relevant for the assessment and decision-making process. Seismic events are classified into four groups: (1) no or minor hazard; (2) seismically-triggered, gravity-driven hazards; (3) stress-adjustment-driven hazards resulting in bulking due to rock mass fracturing; and (4) deformation-driven hazards exploiting existing rock mass damage. Three case histories from 1994–1996, for the 1000 Level South and the 850 Level at Brunswick Mine, are analyzed using this technique to calibrate and verify the proposed methodology.

Key words: Rockbursts, hazard assessment, microseismicity, source parameters.

Introduction

Many research efforts have been directed toward eliminating, mitigating and minimizing rockburst hazard by improved mine design methods, design of energy absorbing or yielding rock support systems, and by better rockburst anticipation techniques (CAMIRO, 1997). Bursting conditions are usually not experienced early in a mine's life; and thus little effort may be placed on preventing burst-prone conditions during mine planning. If problems are encountered later in mine life, it is often not possible to alter the mining method or sequence and ground control engineers may be forced to live with seismicity, requiring procedures to identify potential rockburst hazards and to ensure adequate ground support to minimize risk.

Noranda's Brunswick no. 12 Mine, located in Bathurst, NB, Canada, is a 9000 tonnes per day, zinc-lead-copper-silver operation. Brunswick has a history of

---

[1] Geomechanics Research Centre, Laurentian University, Sudbury, Ontario, P3E 2C6, Canada.
[2] Noranda Mining and Exploration Ltd., Brunswick Mining Division, P.O. Box 3000, Bathurst, New Brunswick, E2A 3Z8, Canada.

microseismicity and has experienced rockburst-related damage to underground excavations. The mine has taken a pro-active approach to mitigating rockburst risk by complementing preventative mine design and ground support initiatives with a ground control program that provides 24-hour access to microseismic monitoring data. Brunswick employs three systems for seismic monitoring: Electrolab MP250 and Queen's Full Waveform (FW) systems for event locations, and an Integrated Seismic System (ISS) for event locations and source parameter information. During the study period (1994–1996), normal daily microseismic activity averaged 400–800 FW system triggers and 20–40 ISS system triggers; however, these numbers could increase tenfold during periods of intense activity. Typically, 75% of these triggers are cultural noise, stemming from ore passes; fill raises as well as development and production blasts (HUDYMA, 1995). Daily data analyses, at the mine, consist of tracking variations in event location clustering and occurrence frequency, and ISS energy index and cumulative apparent volume time histories analyses. These analyses combined with underground observations currently form the basis for workplace closure and re-opening decisions.

Time histories (VAN ASWEGEN and BUTLER, 1993) examine spatial and temporal source parameter variations to monitor rock mass behavior and to predict large magnitude seismic events (potential instabilities). Brunswick has successfully applied this approach, but felt it did not adequately capture or differentiate seismic hazards, largely because seismicity and seismically-induced damge are not restricted to large mangitude events (GIBOWICZ, 1990). HUDYMA (1995) wrote about seismicity at Brunswick, "on an individual basis there is not a good correlation between the size of a seismic event [magnitude] and the level of damage that may be done."

Accepting that current mining conditions and techniques cannot be changed to eliminate rockburst hazards, these hazards must be properly managed as part of a daily ground control decision-making process (e.g., temporary workplace closures and re-openings rehabilitation, support standard revisions). This paper presents a methodology to assess potential rockburst hazards using microseismic source parameters, which is designed to provide a simple but effective means for incorporating the most relevant source parameters into the daily monitoring and decision-making process. The ISS data and observed damage recorded at Brunswick Mine are used to calibrate and verify this methodology.

Rockburst Hazard Assessment

Quantitative seismology (MENDECKI, 1993) can be used to identify stress release (e.g., energy index, seismic energy), stress adjustments (e.g., stress drop or apparent stress) and ground deformation (e.g., apparent volume or seismic moment) indicators which are sensitive to changes in rock mass behavior (stress and strain). By quantifying variations in these parameters within spatial and temporal frames of reference, a Rockburst Hazard Assessment (RHA) methodology was developed. This paper defines a hazard, from a mining perspective, as the potential for damage to an excavation which may impact on operational safety, costs and productivity.

Time histories examine events by cumulative or statistical means to ascertain overall trends, while minimizing fluctuations associated with individual events. The RHA technique uses an alternative approach whereby key events are identified by filtering data using three assessment criteria and then interprets the remaining events by assigning them to one of three characterstic types. The RHA approach is based on a rationale that links localized rock mass behavior, such as seismically-induced falls of ground due to rock mass shaking or bulking due to rock mass fracturing, to specific seismic events. This linkage implicitly assumes that there is a direct correlation between the source parameters of an individual event and the consequences as observed underground. If treated statistically, such linkages would be camouflaged by the averaging process.

The following sections describe the theoretical basis of the methodology and the rationale for selecting energy, apparent stress and seismic moment criteria to describe and differentiate potential rockburst hazards. Ultimately, the goal of the RHA is to arrive at a means to assist the daily decision-making process through definition of operational guidelines to assess progressively worsening and improving conditions in seismically-active workplaces.

Theoretical Considerations

The RHA utilizes three assessment criteria: seismic energy; apparent stress; and seismic moment. These criteria were selected because scalar parameters can be more easily handled and thus lend themselves to routine analyses. Use of source model dependent parameters and seismic moment tensors has been deliberately avoided for this reason. The source parameter calculations employed assume that all events are caused by pure shear failure.

From published seismological relationships (KANAMORI, 1977), it follows that the seismic energy (E) should be proportional to the product of stress drop ($\Delta\sigma$), the co-seismic slip displacement (D) and the area of the fault (A), although the nature of this proportionality depends on the frictional losses during slip.

$$E \propto \Delta\sigma \cdot A \cdot D \quad \text{or} \quad \frac{E}{\Delta\sigma} \propto A \cdot D. \tag{1}$$

All future references to energy (E) refer to seismic energy as it is calculated by the ISS system and is defined in Equation (2) (MENDECKI, 1997).

$$E = E_P + E_{SH} + E_{SV} \tag{2}$$

where P, SH and SV refer to the body wave components.

Similarly, the seismic moment (M_0) is equal to the co-seismic slip displacement (D), the source area (A) and the rigidity or shear modulus (G) of the rock mass containing the seismic source (AKI and RICHARDS, 1980):

$$M_0 = G \cdot A \cdot D. \tag{3}$$

All future references to seismic moment (M_0) refer to seismic moment as it is calculated by the ISS system as an average value from P- and S-wave spectra.

The stress drop ($\Delta\sigma$) is a measure of co-seismic stress adjustment at the source, which is calculated based on a model-dependent source radius (r_0) (BRUNE, 1970; MADARIAGA, 1976). However, the apparent stress (σ_a), which is a measure of the average co-seismic stress adjustment is model independent, and thus a more reliable parameter (WYSS and BRUNE, 1968; GIBOWICZ et al., 1990). For typical conditions encountered in mining situations (URBANCIC et al., 1992) stress drop is proportional to apparent stress such that:

$$\Delta\sigma = \frac{7M_0}{16r_0^3} \propto \sigma_a = G\frac{E}{M_0}. \tag{4}$$

Assuming that apparent stress is proportional to stress drop, the co-seismic deformation (AD) can be described (from Equations (1), (3) and (4)) in terms of energy and apparent stress or the seismic moment:

$$A \cdot D \propto \frac{E}{\sigma_a} \propto M_0. \tag{5}$$

AD is a measure of co-seismic deformation and the volume of rock it affects. In the energy versus seismic moment space (Fig. 1), the co-seismic deformation increases parallel to the $\sigma_a = $ constant line and the co-seismic stress adjustment increases perpendicular to it. If this co-seismic deformation causes rock mass deformation, then the rockburst hazard potential (i.e., the potential for damage to an excavation) is a function of energy, apparent stress and seismic moment. In practice, an increase in either AD or σ_a can be interpreted as an increase in rockburst hazard potential because larger co-seismic deformations strain the rock mass more, or more volume of rock is affected, and larger stress adjustments associated with events of large apparent stress (σ_a) result in stress redistributions that may bring the rock surrounding the event closer to failure. Hence, the rockburst hazard potential should be a function of energy and apparent stress or energy and seismic moment (Equation (5)). The nature of this potential hazard therefore depends on the position of a seismic event in the $E - M_0$ space, which can be tracked using a seismic path concept.

Seismic Path Concept

Stress paths are commonly used in geomechanics to track changes in loading of an element of rock in the principal stress space. Similarly, the authors have considered the variations of energies and seismic moments, in the $\log E - \log M_0$ space, using a seismic path concept. A seismic path consists of events, within a specified volume, plotted in energy-seismic moment space and connected sequentially in time (Fig. 2). A detailed analysis of the data presented on Figure 2 reveals two dominant seismic path trends:

(1)   movement sub-perpendicular to a line of constant σ_a, and
(2)   movement sub-parallel to a line of constant σ_a.

Therefore, based on the methodology presented earlier and reflected in Figure 1, events exceeding a certain σ_a threshold are characteristic of elevated stress adjustments, which may cause damage to excavations in areas of highly stressed rock, and events exceeding a certain M_0 threshold are indicative of large ground deformations, which may cause detrimental rock mass degradation and potential falls of ground.

Assessment Criteria and the Establishment of RHA Thresholds

The RHA methodology presented here utilizes Equation (5) or the three parameters (energy, apparent stress and seismic moment) coupled with the seismic

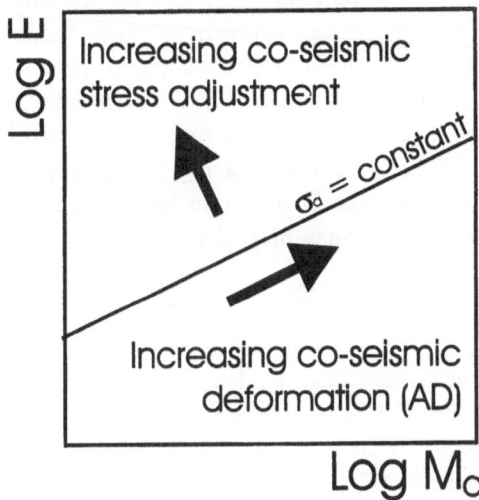

Figure 1
Classification of potential rockburst hazard as stress-adjustment or deformation-driven, based on their position in $\log E - \log M_0$ space.

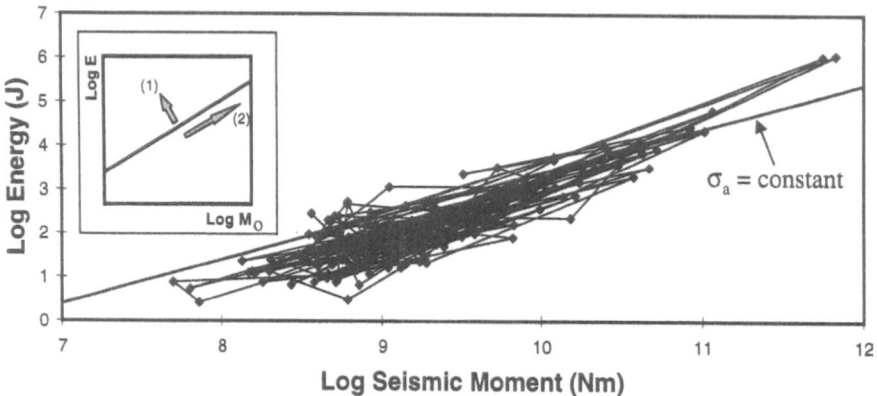

Figure 2
Seismic path for the events recorded during the time period 1–5 December 1994 (1000 Level South).

path trends to assess potential rockburst hazards (i.e., seismically-induced falls of ground, bulking due to rock mass fracturing and deformation-driven hazards). By establishing thresholds for each of these three parameters, the potential hazards can be classified. The assessment criteria should be widely applicable; however, the thresholds employed by the RHA methodology will be site-specific. Each of the assessment criteria is explained in general terms and then selection of its threshold value for the calibration and verification case studies is described. It is anticipated that the criteria thresholds described in this paper would have to be adjusted for different rock mass and stress conditions.

Aside from assessing rockburst hazard potential, the three criteria may also be employed to identify seismic events which have minor hazard potential. If there is insufficient co-seismic deformation or stress adjustment, conditions are not critical (Fig. 3a) (i.e., seismic activity poses only a minor hazard). Events with very low energy content cause little ground motion or low peak particle velocities and thus damage is unlikely. Similarly, an event having a small apparent stress will generate little co-seismic stress adjustment, and fracturing will be unlikely, and small seismic moments will result in minor rock mass deformation.

Energy Criterion

As the energy level increases, so does the ground motion or peak particle velocity (ppv). PERRET (1972) established that energy (E) is proportional to the product of the distance from source (R) and the peak particle velocity (v), and KAISER and MALONEY (1997) developed a scaling law to describe this relationship:

$$\log Rv = 0.5 \log E + 0.40, \tag{6}$$

where Rv is in m^2/s and E is in *MJ*.

KAISER *et al.* (1996) related the ground motion level (ppv) to anticipated rockburst hazards (e.g., falls of ground, bulking due to rock mass fracturing and ejection) and illustrated that as the ppv increases, the potential and type of hazard changes. First, low ground motion levels can trigger seismically-induced falls of ground. As the level increases, bulking due to rock fracturing is encountered, and at even higher levels rock ejection must be anticipated (Fig. 4). Therefore, the energy criterion is selected to characterize the trigger limit for the most prevalent or most critical rockburst hazard at a mine.

Energies recorded at Brunswick vary from 10^{-1} to 10^7 J. Based on this range of energy values, Figure 4 identifies seismically-induced falls of ground as the most likely rockburst hazard. This is consistent with the predominant observed damage mode (falls of ground, back failure and caving). Hence, an energy threshold for triggering of falls of ground must be chosen to cover the most sensitive or critical hazard type (Fig. 3b). Fracturing with rock mass bulking for higher energy events at close distances must also be anticipated as a potential hazard, again consistent with observations. However, this hazard will be assessed by an apparent stress criteria (Fig. 3c). According to Figure 4, rock ejection is an unlikely cause of damage at Brunswick Mine.

BUTLER and VAN ASWEGEN (1993) demonstrated with data from two South African gold mines, that peak particle velocities of less than 0.001 m/s could trigger failures of marginally-stable blocks of rock. At a distance from the source of $R = 100$ m with $v = 0.001$ m/s, Equation (6) yields an energy criterion of 1000 J.

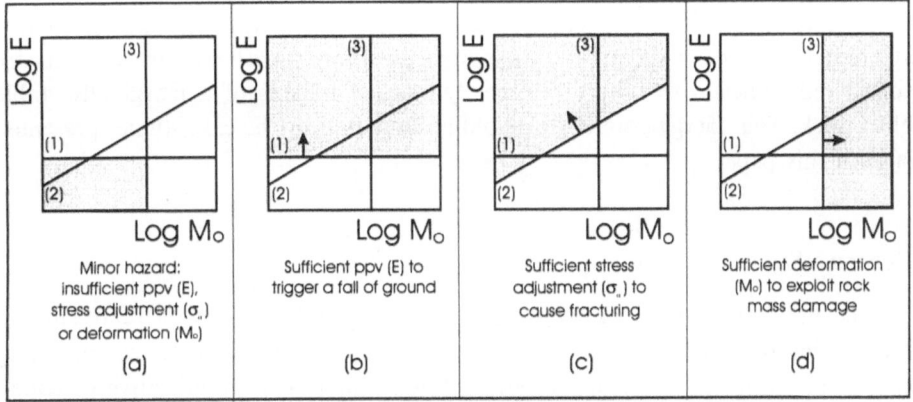

Figure 3
Characterization of (a) minor hazard; (b) triggering falls of ground (FoG) in marginally stable ground; (c) rock mass fracturing; and (d) deformation-induced destabilization of damaged rock mass by three RHA criteria: (1) Energy; (2) Apparent stress; and (3) Seismic moment.

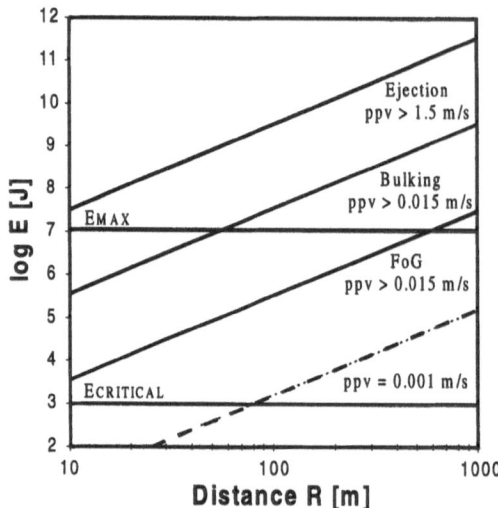

Figure 4
Rockburst hazard as a function of event energy, distance from source and peak particle velocity (after KAISER *et al.*, 1996). Horizontal lines represent the energy criterion ($E_{CRITICAL}$) adopted for Brunswick and the maximum energy (E_{MAX}) recorded at Brunswick.

This value is shown on Figure 4 as $E_{CRITICAL}$. When compared to the trigger limit for falls of ground, shown as 0.015 m/s (KAISER *et al.*, 1996), this threshold should provide a reasonable filter for events that are not likely to cause falls of ground of the type experienced at Brunswick Mine. The energy criterion threshold (1000 J) was established to filter out events that have little impact at the workplace and thus was defined to identify events generating peak particle velocities in excess of 1 mm/s at a design distance of 100 m. From experiences in hard rock mines, these events are considered sufficient to trigger seismically-induced failures of a marginally stable block rock. This energy criteria threshold is supported by the case studies presented later in this paper.

Apparent Stress Criterion

The first trend identified by the seismic path analysis (Fig. 2) represents an increase in apparent stress or co-seismic stress adjustment with little accompanying change in seismic moment or co-seismic deformation. This is indicative of conditions in regions where stresses build-up and suggests an elevated fracturing potential (Fig. 3c). Although, apparent stress alone does not uniquely define an event in the $E - M_0$ space, it does fundamentally relate energy and seismic moment (Equation (4)). Apparent stress is not a measure of absolute stress, however it is a measure of stress change and its magnitude has rockburst and support design implications. As apparent stress increases, larger energy dissipation demands are imposed on the rock and the ground support, and increases beyond its capacity will lead to an enhanced rockburst damage and fracturing potential. Since apparent stress is stress adjustment or stress change and additional increments of stress change lead to rock mass fracture and bulking, deformation of, and imposition of higher load demands on ground support. The impact of an increase in apparent stress on an opening or excavation depends on several factors. In the case of a stable, relatively-undamaged opening, large co-seismic stress adjustments would be required to cause fracturing. Whereas, for openings at depth or in areas of high extraction ratio, which have already experienced rock mass damage and deformation, minor co-seismic stress adjustments may result in damage.

An apparent stress criterion of 7500 Pa was selected as threshold for stress-related damage at Brunswick Mine. The apparent stress threshold was defined based on calibration with field observations to identify larger stress changes or adjustment events, which were considered sufficient to cause fracturing or rock mass bulking. The threshold was calibrated against the larger apparent stress and smaller seismic moment events that were observed on the abutment regions following failure, the migration of seismicity and the redistribution of stresses to those regions. Its effectiveness will be explored in the three case studies presented later.

Seismic Moment Criterion

The second trend, identified by the seismic path analysis (Fig. 2), represents sudden increases in seismic moment or deformation (AD; Equation (3)). These events with large co-seismic deformation strain the rock mass. The impact of these strains depends on the degree to which the rock mass has already been damaged. If cumulative straining or deformation is high then the additional increment of deformation creates conditions conducive to deformation-driven hazards and enhances their potential (Fig. 3d). Aside from these hazards, large seismic moment events are indicative of major shear movements (fault slip) and thus pose an additional rockburst hazard (larger magnitude event).

A seismic moment threshold of 8.8×10^{10} Nm was selected to filter these event types at Brunswick Mine. The seismic moment threshold was defined based on calibration with field observations to identify deformation-driven hazards and was calibrated against the larger seismic moment events that were observed in the stoping area where waste stringer crushing and large-scale failures occurred. Again, this threshold will be evaluated by the following case studies.

Source Parameter Data Errors

The Rockburst Hazard Assessment assesses rock mass behavior relative to critical thresholds of energy, apparent stress and seismic moment. In order to assess individual event source parameter values relative to critical thresholds, one must have a reasonable degree of confidence in the calculated event source parameters. It is recognized that variations in radiation pattern, attenuation and system sensitivity will affect these source parameter values. Even if one assumes an error of one order of magnitude on the calculated source parameter values, this should have little effect on selected energy and seismic moment threshold values, which were defined for energies and seismic moments that varied over eight and seven orders of magnitude, respectively. However, if microseismic data beyond event locations are to be of any use for mining and ground control decision-making, one must be able to use source parameter data either as calculated by the system or as calculated after manual processing, accepting the errors inherent in those calculations.

Assessment Methodology

The three criteria and their thresholds are combined and applied to ISS data within regional polygons to filter out minor hazard potential events, and to classify the remaining events in one of the three categories: (1) seismically-triggered gravity-driven hazard potential, (2) stress-adjustment-driven hazard potential; and (3) deformation-driven hazard potential (Fig. 5a). Events exceeding the energy criterion (E) are considered indicative of a seismically-triggered, gravity-driven hazard potential (e.g., seismically-induced fall of ground). Events exceeding both the energy and the apparent stress criteria (E & σ_a) are considered indicative of a stress-adjustment-driven hazard potential (e.g., bulking due to rock mass fracturing). Events exceeding both the energy and the seismic moment criteria (E & M_0) are considered indicative of a deformation-driven hazard potential (e.g., caving, back failure or large magnitude fault-slip rockburst). Events meeting or exceeding these criteria are plotted on Log Energy versus Time plots. The criteria employed, and symbols used to represent them, are shown in Figure 5a and a sample Log Energy versus Time plot is shown in Figure 5b. Blast times are shown by plus signs above the time axis.

These plots are analyzed for precursory, failure and decay trends correlated to observed damage (shown as vertical dashed lines; Fig. 5b). Precursory trends are defined as warning signals consisting of a sequential spatial and temporal build-ups of indicators (i.e., $E \Rightarrow E$ & $\sigma_a \Rightarrow E$ & M_0). For example, first exceeding a critical energy level, then a critical energy and apparent stress, and eventually, a critical energy and moment threshold; in short $E \Rightarrow E$ & $\sigma_a \Rightarrow E$ & M_0. These trends identify the area as a region of elevated rockburst hazard potential, and indicate a progressive worsening of conditions and potential hazards from shakedown (E) to fracturing (E & σ_a) to deformation (E & M_0) resulting in the exploitation rock mass damage. Conversely, decay trends refer to a progressive improvement of conditions represented by the sequential disappearance indicators (i.e., E & $M_0 \Rightarrow E$ & $\sigma_a \Rightarrow E$). Events corresponding to these trends are then plotted on plans and sections for further analysis to identify regions of elevated rockburst hazard potential.

The usefulness of the RHA is demonstrated in its application to three case histories, which consider two mining blocks and several time periods.

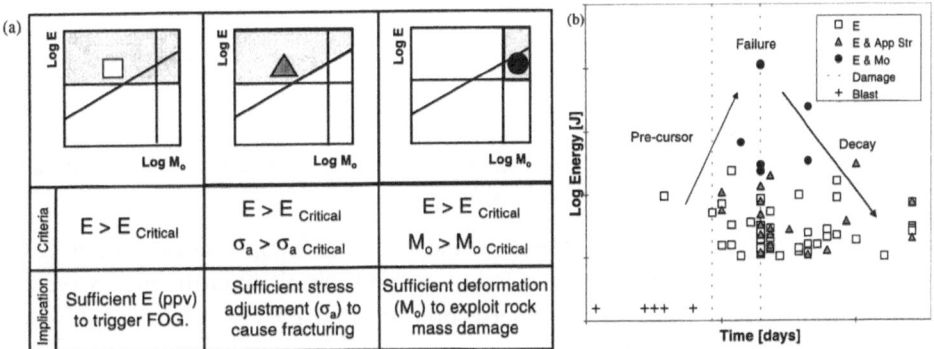

Figure 5

(a) Characterization of microseismic events: exceeding the energy criterion ($E > E_{Critical}$) as indicative of a potential for seismically-triggered, gravity-driven hazards; exceeding the energy and apparent stress criteria ($E > E_{Critical}$ and $\sigma_a > \sigma_{aCritical}$) as indicative of a potential for stress-adjustment driven hazards; or exceeding the energy and seismic moment criteria ($E > E_{Critical}$ and $M_0 > M_{0Critical}$) as indicative of a potential for deformation-driven hazards. (b) Sample hazard assessment showing a precursory sequential build-up of indicators ($E \Rightarrow E$ & $\sigma_a \Rightarrow E$ & M_0) prior to a failure (indicated by the dashed line) and followed by a sequential decay of indicators (E & $M_0 \Rightarrow E$ & $\sigma_a \Rightarrow E$).

Case Histories

Brunswick operates on three main production levels (725, 850 and 1000 meter levels). Three case histories are examined in this paper, two from 1000 Level South and one from the 850 Level (Fig. 6). Constant criteria thresholds (i.e., $E = 1000$ J, $\sigma_a = 7500$ Pa, $M_0 = 8.8 \times 10^{10}$ Nm) were employed in all three analyses. These thresholds were first selected empirically for one part of Brunswick Mine applying source parameter data and observed damage, and then calibrated for other parts of the mine.

Geologically, the 1000 Level South consists of sulphide ore interfingered with waste meta-sediments and a porphyry dyke, and bounded by footwall and hanging-wall sediments. Traditionally, ground control concerns have been attributed to the

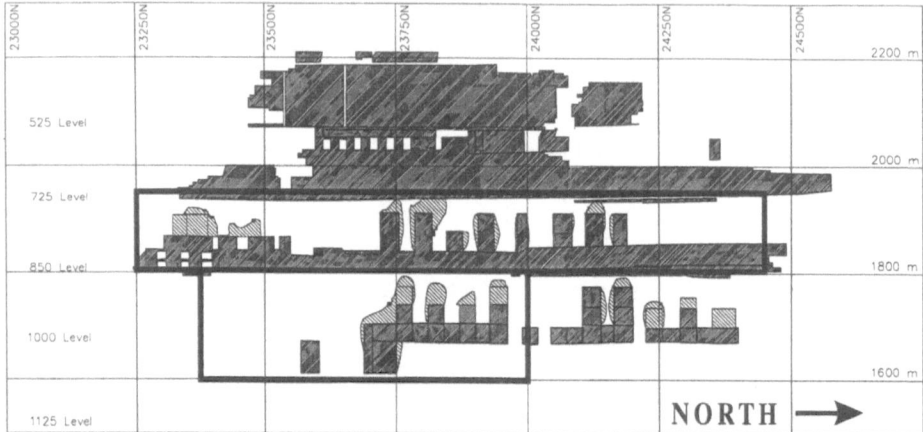

Figure 6
Vertical long-section of the Main Ore Zone (MOZ), looking West, indicating mining blocks. Stope caving and overbreak are indicated by lighter hatching. The rectangles indicate the 1000 Level South and 850 Level study areas.

contrast between weaker meta-sediments and the more competent ore, resulting in squeezing of the meta-sediments and fracturing and seismicity in the ore. A typical fall of ground type failure at an intersection affected by a meta-sediment waste stringer is shown in Figure 7.

Figure 7
Intersection in 238–7 Cross-Cut (X/C) (1000 Level South) following failure along the waste stringer.

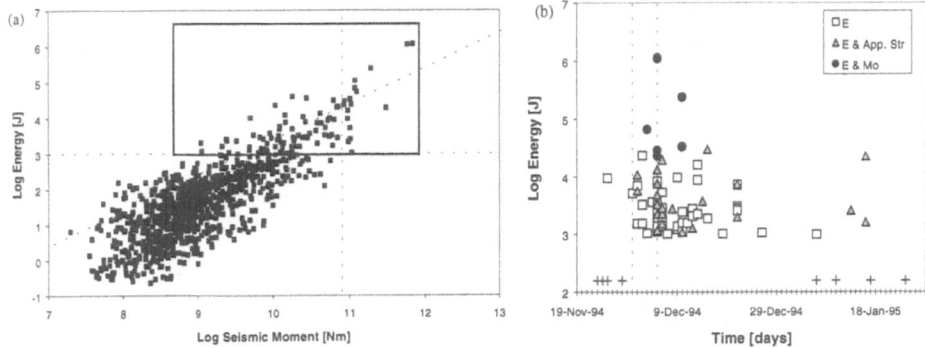

Figure 8

(a) Assessment criteria applied to all events; the rectangle highlights events meeting the criteria. (b) rockburst hazard assessment; dashed lines indicate failures and crosses indicate blasts.

The geology of the 850 block is similar to the 1000 Level. Ground control concerns are related to stope backs caving into the sill and hourglassing of stope sidewalls due to the creation of unfavorable mining geometries (i.e., slender pillar geometries for the secondary stopes).

Case 1: 1000 Level South—September 1994 to January 1995

A total of 489 events from the time period 25 November 1994–25 January 1995 were analyzed. The events were filtered by applying the three criteria introduced earlier: energy (1000 J), seismic moment (8.8×10^{10} Nm) and apparent stress (7500 Pa)—Figure 8a. This filtering identified 70 events which met at least the energy criterion. The assessment highlights a period of major activity associated with blast-triggered waste stringer mobilization (crushing) (Fig. 8b). Blast and failure locations and dominant geological structure (waste stringer) discussed later are shown in Figure 9a. Figure 9b illustrates all 1000 Level South events recorded during this period.

Two failures occurred during this period. A fall of ground in 336–8[3] cross-cut (X/C) and continued pillar deterioration in 236–8 occurred on 30 November, 1994 (indicated by the first dashed line in Fig. 8b). Two E & M_0 indicators were recorded, one 5 days prior to the failure and another on the day of the failure. Neither of which was consistent with the failure location. The subsequent failures, 335–8 Stope Back, continued failure of 336–8 X/C and significant ground deterioration along 34–8 through 39–8 Stopes on 2 and 3 sub-levels, occurred on 5 December 1994 (second dashed line in Fig. 8b). In the five days prior to these failures, 10 indicators ($6E$, $2E$ & σ_a and $2E$ & M_0) were recorded. These indicators

---

[3] First digit refers to sub-level (omitted if no reference to sub-level made); the second and third digits are the stope name; and the hyphenated number refers to the ore lense.

corresponded to a systematic build-up of indicators or precursor (i.e., $E \Rightarrow E$ & $\sigma_a \Rightarrow E$ & M_0) and were consistent with the subsequent failure locations, delineating these areas as regions of elevated rockburst hazard potential (Fig. 10a).

On 5 December 1994, the day of the failures, 20 indicators were recorded ($10E$, $6E$ & σ_a, $4E$ & M_0). These event locations are consistent with the significant ground deterioration observed between the 34–8 and 39–8 stopes on #2 and 3 sub-levels (Fig. 10b), and with the waste stringer (HW sediments and tuffs) orientation (Fig. 9a).

Following the failures, December 6 to 26, 1994, there was a systematic decay of indicators (E & $M_0 \Rightarrow E$ & $\sigma_a \Rightarrow E$) (Fig. 10c). E & M_0 indicators decayed first over 9 days, then the E & σ_a events decayed over 16 days and E only events decayed over 21 days. These indicator locations corresponded to the decay of hazard indicators in the failed region and a migration of seismicity along the waste stringer to the adjacent abutments. The large deformation criteria events (E & M_0), located along the Waste Stringer (Fig. 10c) are indicative of a deformation-driven hazard associated with crushing of the stringer and seismic migration to adjacent abutments (Figs. 10c and 8b). Interestingly, events located on the South Regional Abutment (Fig. 10c) satisfy the E & σ_a criteria; indicating loading without much accompanying deformation. This is indicative of a potential for a stress-adjustment-driven or fracturing hazard rather than a deformation-driven one, and implies a need for energy absorption capacity in HW access support. The rockburst hazard assessment technique significantly reduces data (to 14% in this case) for high activity periods while focusing attention on key events, which clearly pinpoint areas of elevated risk, allowing the ground control personnel to take appropriate actions.

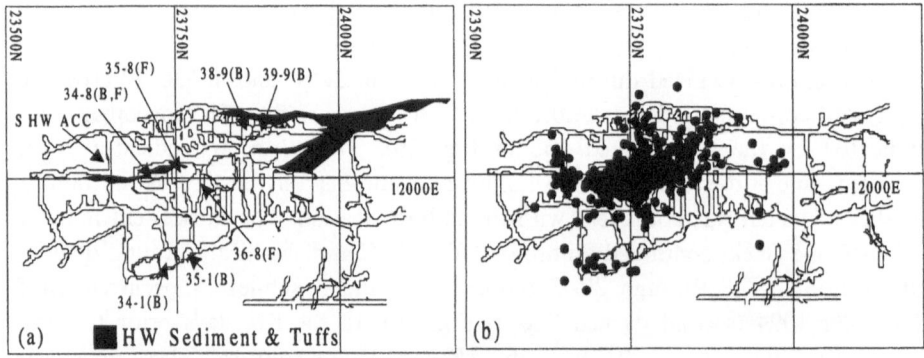

Figure 9
1000 Level South #2 Sub-level plan (a) Blasting (B), failure (F) and waste stringer locations. (b) All recorded events (25 November 1994–26 December 1994).

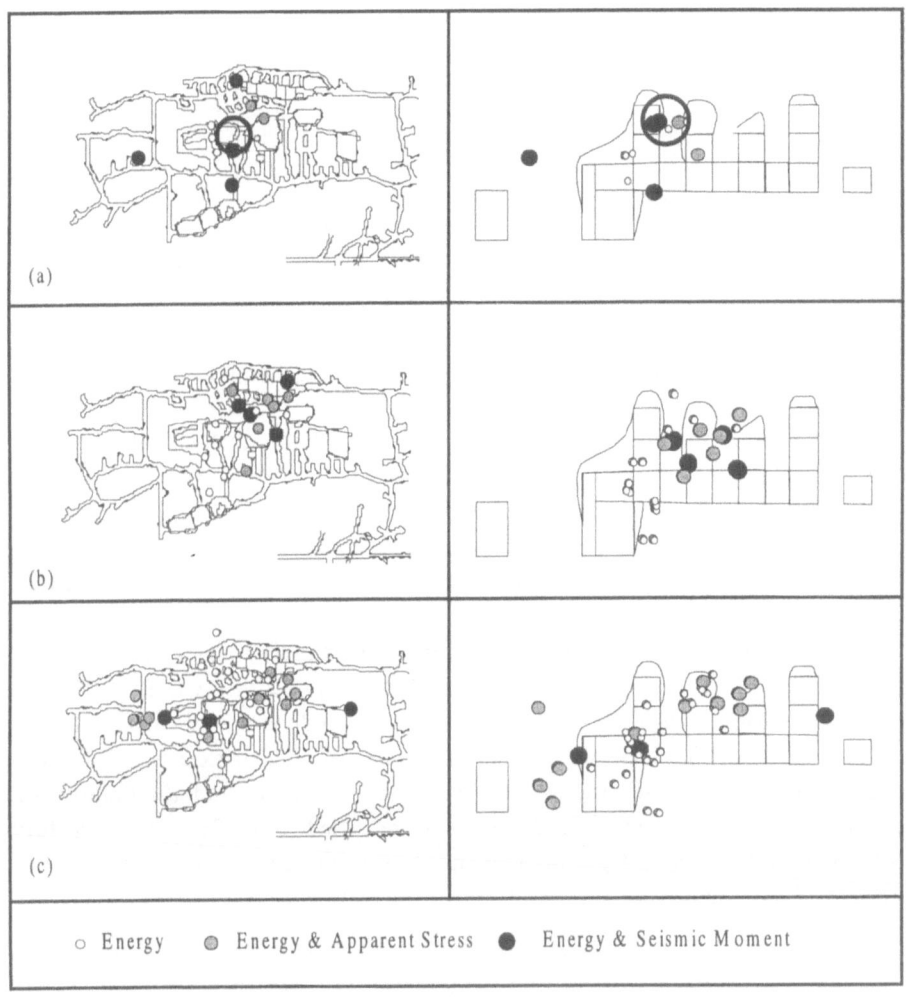

Figure 10
Rockburst hazard assessment indicator event locations shown on #2 Sub-level plan and Main Ore Zone
long section. (a) 25 Nov. 1994–4 Dec. 1994. (b) 5 December 1994. (c) 6 Dec. 1994–26 Dec. 1994. The
thick black circle indicates the location of the 5 December 1994 failures.

Case 2: 1000 Level South—September 1995 to January 1996

A total of 904 events recorded from 1 October 1995 to 31 January 1996 was
analyzed. The events were filtered by applying the same criteria thresholds as
before. Filtering results in 103 events, which met at least the energy criterion. Three
failures occurred during this period. Blasting and failure locations and critical
geology (waste stringer) referenced in the following discussion are shown in Figure
11.

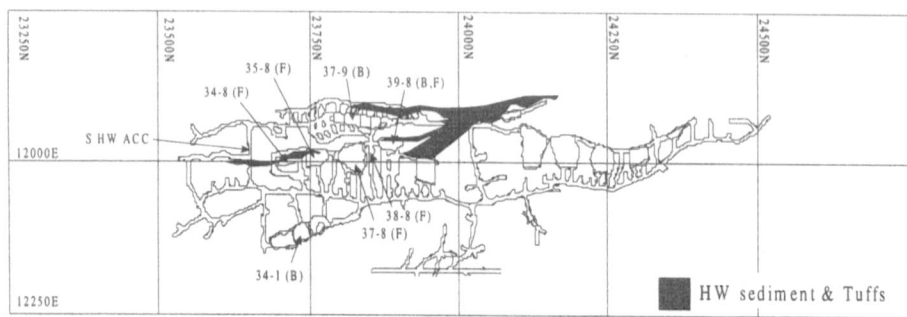

Figure 11
1000 Level #2 Sub-level plan blasting (B), failure (F) and waste stringer locations shown.

In the three days prior to a 300 tonne rockburst in the back of 239–8, 7 indicators were recorded ($5E$, $1E$ & σ_a and $1E$ & M_0). On 29 November 1995, the day of the rockburst, 2 indicators were recorded ($1E$ and $1E$ & M_0). One energy indicator was recorded 4 days after the failure. Precursor, failure and decay events correspond to and clearly delineate the rockburst location as an area of elevated rockburst hazard potential (Figs. 12a,b).

There were few precursory indicators in the 5 days prior to 19 December 1995 failures (237–8 Back Failure, 139–8 Back Failure, 238–8 HW ACC Back Failure and 235–8 continued Back and HW Failure). Only 3 indicators ($1E$ and $2E$ & M_0) were recorded, one of which was located in the area of the subsequent 238–8 HW Access Back Failure location (Fig. 13a). However, on the day of the failures, 41 indicators ($22E$, $7E$ & σ_a and $12E$ & M_0) were recorded (Fig. 13b). The indicators recorded between 8 am–12 pm on December 19 exhibited the typical precursory sequence ($E \Rightarrow E$ & $\sigma_a \Rightarrow E$ & M_0) (Fig. 14). Many of these event locations

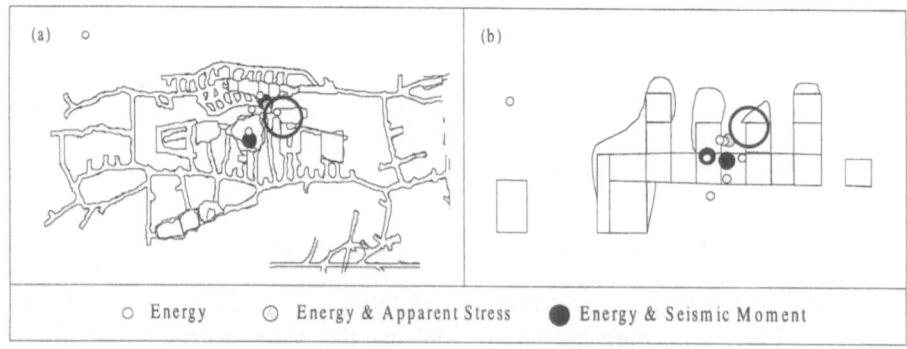

Figure 12
1000 Level South rockburst hazard assessment (26 November 1995–3 December 1995) indicator locations shown on (a) #2 Sub-level plan and (b) Main Ore Zone long section. The thick black circle indicates the location of the 28 November 1995 rockburst.

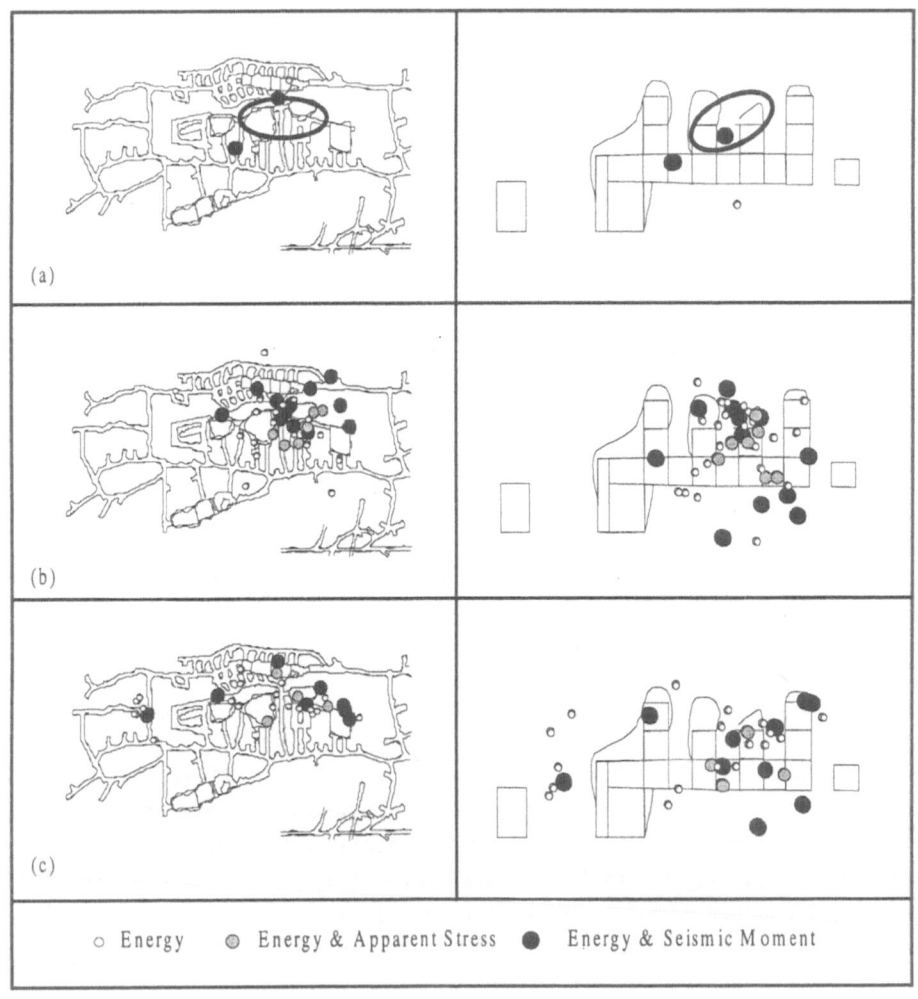

Figure 13
Rockburst hazard assessment indicator event locations shown on #2 Sub-level plan and Main Ore Zone long section. (a) 14 Dec. 1995–18 Dec. 1995, (b) 19 Dec. 1995, (c) 20 Dec. 1995–6 Jan 1996. The thick black ellipse indicates the location of the 19 December 1995 failures.

corresponded to the waste stringer and were located in the affected failure areas, identifying these areas as having an elevated rockburst potential (Figure 13b).

There was a post-failure decay period (20 December–6 January). During this period, events were primarily constrained to the waste stringer and there was again a visible migration of seismicity to both abutment regions (Fig. 13c). The decay pattern was as follows: E & σ_a over 2 days, E & M_0 over 10 days; and E over 17 days.

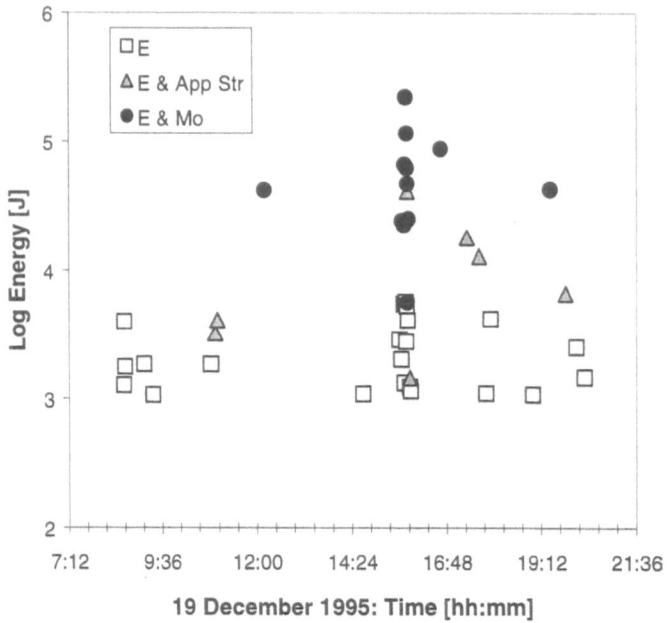

Figure 14
Hazard assessment indicator sequence on 19 December 1995, the day of the failures.

Between 2–8 January 1996, a fall of ground occurred in 334–8; however, there were no indicators directly associated with this failure location and time. The possibility that the failure could have been triggered by an event below the energy criterion was investigated (note that this area was affected by large deformation events in December 1994 (Fig. 10c)). All recorded events within 100 m of the 334–8 stope were examined in order to identify potential fall of ground trigger events,

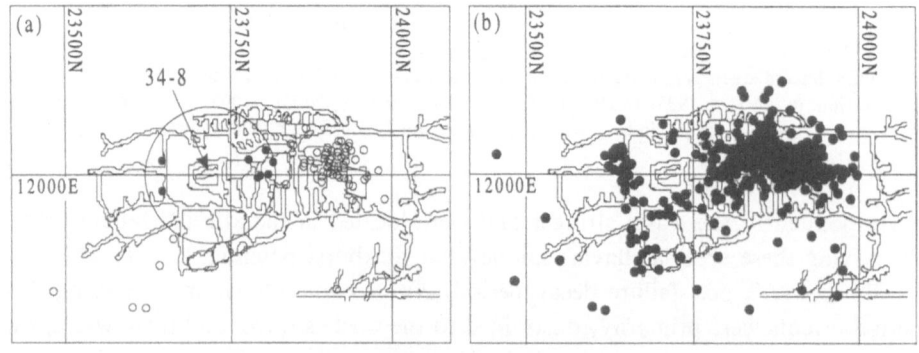

Figure 15
All events (a) 2–8 January 1996 and (b) 14 December 1995–6 January 1996 plotted on the #2 Sub-level plan.

which would have been filtered out by the energy criterion (Fig. 15a). The identified events caused ppv(s) on the order of 0.1 mm/s. During the entire analysis period no ISS events were recorded in 34–8 (Fig. 15b); however, seismicity was recorded in the surrounding stopes, despite the fact that the waste stringer was being mobilized and was present within the stope. Based on the seismic evidence, 34–8 is interpreted as a failure due to gradual rock mass which yielded around the stope causing stress relaxation, which promoted gravity-driven failures. This failure in 334–8 was most likely not seismically-triggered by an event occurring in the 1000 Level South.

Case 3: 850 Level South—September 1994 to January 1995

A total of 578 events recorded from 21 September 1994 to 25 January 1995 were analyzed. The events were filtered by applying the same criteria thresholds as before. Filtering results in 40 events, which met at least the energy criterion. Two failures were recorded during this period. Blasting and failure locations discussed are shown in Figure 16.

On 9 December 1994, there was a wall failure in the 285–3 X/C. One precursory energy indicator was consistent with the location of the subsequent failure (Fig. 17a). During the failure, 3 indicators (1E and 2E & σ_a) were recorded. These indicators delineated the 285–3 X/C failure location (Fig. 17b), an area of elevated rockburst hazard potential. Failure and decay intervals run concurrently. The failure was both small, 2–3 tons, localized, occurring along a waste sediment/dyke contact, and resulted in neither seismic migration nor subsequent damage, suggesting that little stress adjustment was involved in this failure.

On 10 December 1994 failures were recorded in 279–7 Cemented Rockfill (CRF) Wall Failure/379–7 Hangingwall (HW) Failure. No indicators were consis-

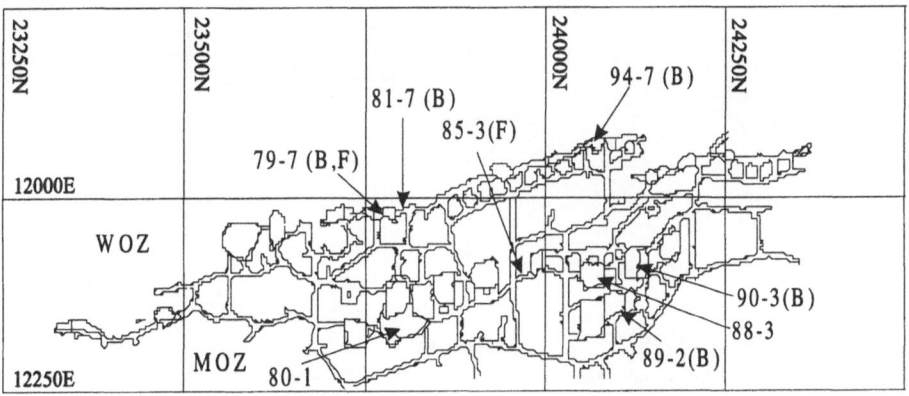

Figure 16
850 Level #4 Sub-level plan blasting (B) and failure (F) locations shown.

tent with these failure locations (Fig. 17a) and neither the 379–7 HW nor the 279–7 CRF locations were identified as areas of elevated rockburst hazard potential. The 279–7 CRF failure was unlikely to have generated seismicity, although both failures could potentially have been seismically-triggered (e.g., shakedown failures). The possibility of seismically-induced failure triggered by an event below the energy criterion was again investigated. Inscribing a sphere of 100 m about 279–7/379–7 stopes, two events fell within the 100 m radius (Fig. 18a). The most likely trigger event occurred on December 10 (the failure date) at a distance of 56 m from the failure location, with an energy of 53.1 J, which translates to a ppv of 0.33 mm/s (Equation (6)). This event cannot conclusively be ruled out as a potential trigger for seismically-induced failure. Nevertheless, this example suggests that the rock mass was sufficiently damaged or deformed and that rather minor shaking could cause gravity-driven instabilities.

From 19 December 1994 to 19 January 1995, there were no recorded failures. However, build-up and decay indicator trends were observed (i.e., $E \Rightarrow E$ & $M_0 \Rightarrow E$). Indicator locations correspond to a band of pillars between 280–1 and 288–3 stopes (Fig. 17c), and identify them as areas of elevated rockburst hazard potential. E & M_0 indicators (deformation) correspond to these pillar locations and locate along a waste (HW sediments and tuffs)/ore contact. Therefore, these events may be indicative of pillar crushing or waste stringer crushing inside the pillar or in the footing of the pillar.

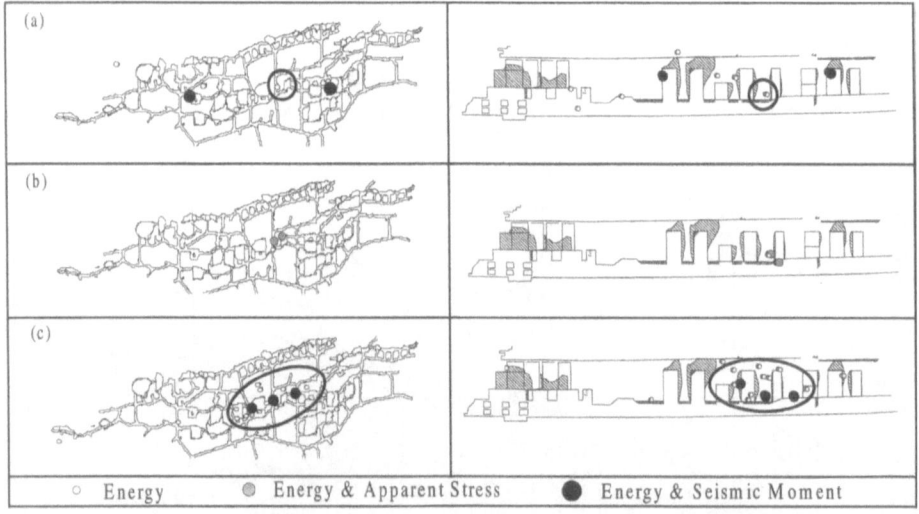

Figure 17
Rockburst hazard assessment indicator event locations shown on #4 Sub-level plan and Main Ore Zone long section. (a) 24 Nov. 1994–8 Dec. 1994. The thick black circle indicates the location of the 9 December 1994 failure. (b) 9–11 Dec. 1994. (c) 19 Dec. 1994–19 Jan. 1995. The thick black ellipse indicates the location of the secondary stope pillars.

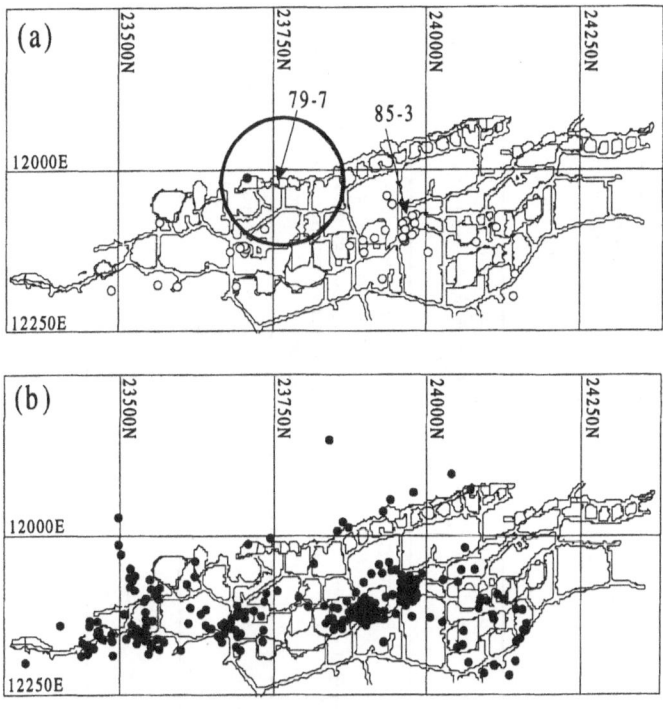

Figure 18
All 850 Level events (a) 8–11 December 1994; the large circle indicates potential trigger events for the 279–7/379–7 failures and the black dot indicates the most likely trigger event, (b) 24 November 1994–19 January 1995 plotted on the #4 Sub-level plan.

Figure 18b shows all 850 Level events recorded from 24 November 1994 to 19 January 1995. In total, 283 events were recorded of which 28 events met indicator criteria (10%).

Summary

The observed damage, the RHA indicators (precursory, failure and decay), the indicator sequences and the success of the RHA methodology, for the three case studies, are summarized in Table 1. Broadly, the failures analyzed can be classified as instances:

(1) where seismicity leads to failure (e.g., rockburst damage to excavations) and the seismicity can be used to anticipate the failure and its precursor (e.g., the development of a state of elevated hazard potential); or

Table 1

Summary of Rockburst Hazard Assessment case studies.

Case	Damage	Indicators	Sequence	Notes
1	30 November 1994 • 336–8 X/C fall of ground • 236–8 pillar failure	Precursor: 1 E & M_0 Failure: 1 E & M_0 Decay: n/a	n/a	Indicators were not consistent with failure locations.
1	5 December 1994 • 335–8 back failure • 336–8 X/C failure • Significant ground deterioration 34–8 to 39–8 stopes on #2 & 3 sub-levels	Precursor: 6 E, 2 E & σ_a, 2 E & M_0 Failure: 10 E, 6 E & σ_a, 4 E & M_0 over 9 days, E & σ_a over 16 days, E over 21 days Decay: E & $M_0 \Rightarrow E$ & $\sigma_a \Rightarrow E$	Precursor: $E \Rightarrow E$ & $\sigma_a \Rightarrow E$ & M_0 Decay: E & $M_0 \Rightarrow E$ & $\sigma_a \Rightarrow E$	Precursory (5 days prior to) and failure indicators were consistent with the failure locations. Decay events showed migration of seismicity along the waste stringer to the abutments.
2	28 November 1994 • Rockburst 239–8	Precursor: 5 E, 1 E & σ_a, 1 E & M_0 Failure: 1 E, 1 E & M_0 Decay: E over 1 day	Precursor: $E \Rightarrow E$ & $\sigma_a \Rightarrow E$ & M_0 Decay: E	Precursory (3 days prior to) and failure indicators were consistent with the failure location.
2	19 December 1995 • 237–8 back failure • 139–8 back failure • 238–8 HW ACC failure • 235–8 back and HW failures	Precursor: 1 E, 2 E & M_0 Failure: 22 E, 7 E & σ_a, 12 E & M_0 Decay: E & σ_a over 2 days, E & M_0 over 10 days, E over 17 days	Precursor: $E \Rightarrow E$ & $\sigma_a \Rightarrow E$ & M_0 Decay: E & $\sigma_a \Rightarrow E$ & $M_0 \Rightarrow E$	Precursory (hours prior to) and failure indicators were consistent with the failure locations. Decay events showed migration of seismicity along the waste stringer to the abutments.
2	2–8 January 1995 • 334–8 fall of ground	No indicators	n/a	The failure was believed to have been the result of rock mass degradation leading to loss of self-supporting capacity.
3	9 December 1994 • 285–3 X/C wall failure	Precursor: 1 E Failure: 1 E & σ_a Decay: $E \sigma_a$ and E over 2 d	Precursor: $E \Rightarrow E$ & $\sigma_a \Rightarrow E$ Decay: E & $\sigma_a \Rightarrow E$	Precursory, failure and decay indicators were consistent with the failure location. Limited localized damage.
3	10 December 1994 • 279–7 CRC failure • 379–8 HW failure	No indicators	n/a	The failure was believed to have been the result of rock mass degradation leading to loss of self-supporting capacity.
3	19 December 1994–19 January 1994 • No recorded damage but precursory and decay trends corresponding to unmined secondary stopes were observed.	Multiple E and E & M_0 indicators over a month long period.	Precursor: $E \Rightarrow E$ & M_0 Decay: E	The large deformation events (E & M_0) were consistent with the pillar locations, possibly indicative of crushing of the waste stringer within the pillars.

(2) where previous seismicity has resulted in rock mass degradation and the failure occurs as the result of the degradation as opposed to seismicity recorded at the time of the failure. In these cases the seismicity is not likely useful for anticipating the failure or its precursor.

Of the seven failures analyzed, five fall into the first category, and in four of the five cases the failure location was successfully identified by the RHA as an area of elevated rockburst hazard potential. The remaining two cases fall into the second category, where the failures appeared to be the result of progressive rock mass degradation leading to the loss of the rock's self-support capacity and resulting in gravity-driven failures. As such these failures were not identified as areas of elevated rockburst hazard potential by the RHA assessment.

The RHA identifies both precursory (i.e., $E \Rightarrow E$ & $\sigma_a \Rightarrow E$ & M_0) and decay (i.e., E & $M_0 \Rightarrow E$ & $\sigma_a \Rightarrow E$) trends, which are indicative of a progressive worsening and improving, respectively, of conditions in seismically active areas. These trends and their indicators vary as a function of level of rock mass degradation. Typically, as degradation accumulates, the rock mass loses its capacity for stress transfer (e.g., stress redistribution and the migration of seismicity). This is captured by the absence of E & σ_a indicators in both the precursory and decay sequences (i.e., $E \Rightarrow E$ & $M_0 \Rightarrow E$). In addition to the absence of E & σ_a indicators, as rock mass degradation accumulates, the duration of both the precursory (i.e., from days to hours prior to failure) and decay intervals (i.e., from weeks where stress redistribution and migration occur to days where they do not) are shortened.

The ability to identify areas of elevated rockburst hazard potential, coupled with rock mass condition dependent precursory and decay trends, makes the Rockburst Hazard Assessment ideally suited as a basis for closure/reopening and support requirement decision-making in seismically-active workplaces.

Conclusions

A rockburst hazard assessment technique is presented that is able to successfully delineate areas of elevated rockburst hazard potential, and which identifies precursory ($E \Rightarrow E$ & $\sigma_a \Rightarrow E$ & M_0) and decay trends (E & $M_0 \Rightarrow E$ & $\sigma_a \Rightarrow E$) indicative of progressive worsening and improving conditions in seismically-active workplaces, at Brunswick Mine. The RHA establishes source parameter performance criteria and thresholds which can be (and have been) easily incorporated into the daily ground control decision-making process. This technique is intended for use with existing time history and event location analysis techniques. Linking hazard indicators with event locations provides a means to delineate indicators by mechanism; e.g., stress-adjustment-driven/abutment failures characterized by E & σ_a indicators, or deformation-driven/waste stringer instability defined by E & M_0 indicators. The three cases presented illustrate the robustness and applicability of unique threshold criteria to different mining blocks. The same threshold values were shown to be applicable for the three case study areas, however different thresholds may have to be established for other mines or rock mass conditions. Differences in precursor and decay trends highlight variations in rock mass condition and failure mechanisms. This assessment procedure provides the basis for assessing support requirement because areas affected by critical seismicity can be identified. It also assists in defining a workplace closure/reopening policy based on spatial and temporal precursory trends and decay intervals. One caution is offered. An assessment of this nature places increased emphasis on source parameters; the quality of which relies on the quality of the acquisition and the data processing system. Lack of high quality data may lead to erroneous interpretations.

Acknowledgements

The authors would like to acknowledge Brunswick Mine and the Noranda Technology Centre for their financial support and assistance. This manuscript benefited from the helpful comments and suggestions of Dr. S. Talebi and two anonymous reviewers.

REFERENCES

AKI, K., and RICHARDS, P. G., *Quantitative Seismology: Theory and Methods*, Vol. 1 (W. H. Freeman, San Francisco 1980).

BRUNE, J. N. (1970), *Tectonic Stress and Spectra of Seismic Shear Waves from Earthquakes*, J. Geophys. Res. *75*, 4997–5009.

BUTLER, A. G., and VAN ASWEGEN, G., *Ground velocity relationships based on a large sample of underground measurements in two South African mining regions.* In *Proc. 3rd Int. Symp. Rockbursts and Seismicity in Mines* (A. A. Balkema, Rotterdam 1993) pp. 41–48.

CAMIRO, *Canadian Rockburst Research Program 1990–95*, Vols. 1 and 2 (CAMIRO Mining Division, Sudbury 1997).

GIBOWICZ, S. J. (1990), *Keynote Lecture: The mechanism of seismic events induced by mining: A review.* In *Proc. 2nd Int. Symp. Rockbursts and Seismicity in Mines* (A. A. Balkema, Rotterdam 1990) pp. 3–27.

GIBOWICZ, S. J., HARJES, H. P., and SCHAFER, M. (1990), *Source Parameters of Seismic Events at Heinrich Robert Mine, Ruhr Basin, Fedral Republic of Germany: Evidence for Non-double-couple Events*, Bull. Seismol. Soc. *80*, 88–109.

HUDYMA, M. (1995), *Seismicity at Brunswick Mining.* In *Proc. of the Quebec Mining Association Ground Control Colloque*, Val d'Or, Quebec, Canada, 18 pp.

KAISER, P. K., and MALONEY, S. M. (1997), *Scaling Laws for the Design of Rock Support*, Pure appl. geophys. *150*, 415–434.

KAISER, P. K., MCCREATH, D. R., and TANNANT, D. D., *Canadian Rockburst Support Handbook* (Geomechanics Research Centre, Laurentian University, Sudbury 1996).

KANAMORI, N. (1977), *The Energy Release in Great Earthquakes*, J. Geophys. Res. *82*(20), 2981–2987.

MADARIAGA, R. (1976), *Dynamics of an Expanding Circular Fault*, Bull. Seismol. Soc. *66*, 639–666.

MENDECKI, A. J., *Real time quantitative seismology in mines*. In *Proc. 3rd Int. Symp. Rockbursts and Seismicity in Mines* (A. A. Balkema, Rotterdam 1993) pp. 287–295.

MENDECKI, A. J. (ed.), *Seismic Monitoring in Mines* (Chapman and Hall, London 1997).

PERRET, W. R. (1972), *Seismic-source Energies of Underground Nuclear Explosions*, Bull. Seismol. Soc. *62*(3), 763–774.

URBANCIC, T. I., YOUNG, R. P., BIRD, S., and BAWDEN, W. B. (1992), *Microseismic source parameters and their use in characterizing rockmass behaviour: considerations from Strathcona Mine*. In *Proc. of the 94th Annual General Meeting of the CIM*, Montreal, 36–47.

VAN ASWEGEN, G., and BUTLER, A. G., *Applications of quantitative seismology in South African gold mines*. In *Proc. 3rd Int. Symp. Rockbursts and Seismicity in Mines* (A. A. BALKEMA, Rotterdam 1993) pp. 41–48.

WYSS, M., and BRUNE, J. N. (1968), *Seismic Moment, Stress and Source Dimension for Earthquakes in the California-Nevada Region*, J. Geophys. Res. *73*, 4681–4694.

(Received February 9, 1998, accepted August 7, 1998)

Pure appl. geophys. 153 (1998) 67–92
0033–4553/98/010067–26 $ 1.50 + 0.20/0

Pure and Applied Geophysics

A Tensile Model for the Interpretation of Microseismic Events near Underground Openings

Ming Cai,[1] Peter K. Kaiser[1] and C. Derek Martin[1]

Abstract—For small-scale microseismic events, the source sizes provided by shear models are unrealistically large when compared to visual observations of rock fractures near underground openings. A detailed analysis of the energy components in data from a mine-by experiment and from some mines showed that there is a depletion of S-wave energy for events close to the excavations, indicating that tensile cracking is the dominant mechanism in these microseismic events.

In the present study, a method is proposed to estimate the fracture size from microseismic measurements. The method assumes tensile cracking as the dominant fracture mechanism for brittle rocks under compressive loads and relates the fracture size to the measured microseismic energy. With the proposed method, more meaningful physical fracture sizes can be obtained and this is demonstrated by an example on data from an underground excavation with detailed, high-quality microseismic records.

Key words: Microseismic event, seismic energy, source size, stress-induced fracturing, tensile fracturing.

Introduction

The concept of deep geological disposal of high-level radioactive waste is based on isolation of the waste from the biosphere by multiple barriers. One of these barriers is the host rock for the repository which must remain mechanically stable for long times. However, construction of a repository at depths of 500 to 1000 m will require the excavation of access tunnels, shafts and rooms. Excavating any of these underground openings in a brittle rock mass at depth results in the formation of fractures in the vicinity of the opening. The extent of this fracturing depends on several factors; the most significant being the stress to strength ratio near the boundary of the opening. The nuclear waste industry has carried out several large-scale *in situ* experiments over the past 15 years to assess the extent of fracturing that could form around openings in brittle rocks (OLSSON and WIN-BERG, 1996). Two of these more recent experiments (AECL's Mine-by Experiment

[1] Geomechanics Research Centre, Laurentian University, Sudbury, Ontario, P3E 2C6, Canada.

at the Underground Research Laboratory (URL) and SKB's Zebex Experiment) used microseismic monitoring to quantify the fracturing (e.g., TALEBI and YOUNG, 1992; MARTIN and READ, 1996; FALLS and YOUNG, 1996).

AECL's Mine-by Experiment was designed to monitor the excavation-induced response around a 3.5-m-diameter test tunnel excavated without explosives in unfractured Lac du Bonnet granite (MARTIN and READ, 1996). The test tunnel was excavated over a six month period and a well-developed notch formed in the roof and floor as the tunnel was excavated. The method of investigation was chosen to provide physical observations of the failure process (MARTIN *et al.*, 1997). Upon completion of the test tunnel a detailed investigation was carried out regarding the amount of physical damage that could be observed by carefully excavating observation slots in the roof and floor of the tunnel (Fig. 1). In addition to the monitoring of displacements, the fracturing process was also captured by a full wave-form microseismic system consisting of sixteen triaxial accelerometers. The accelerometer array was designed for focal sphere coverage and a source location accuracy of about ± 0.25 m near the center of the tunnel (TALEBI and YOUNG, 1992; YOUNG, 1993).

Figure 1
Example of an observation slot used to observe the rock-mass quality below the well-developed notch in the floor of the Mine-by test tunnel. There was no physical evidence of fractures observed outside the boundary of the notch.

The application of acoustic emission/microseismic (AE/MS) monitoring to investigate stress-induced fracturing is not new and the techniques have been used extensively to source locate brittle fracturing in the laboratory and in deep mines (e.g., LOCKNER et al., 1992; ECOBICHON et al., 1992). However, AECL's Mine-by Experiment and the preliminary shaft monitoring experiment (GIBOWICZ et al., 1991; TALEBI and YOUNG, 1989, 1992) are, to the authors' knowledge, the only experiment carried out in unfractured massive granite with a microseismic monitoring system designed to capture the full wave-form seismic signals generated by the excavation-induced fracturing. The focus of this work was to provide information on the fundamental mechanisms which caused the fracturing in the vicinity of the opening and the physical dimensions of these fractures. This information is deemed necessary to quantify the degree of fracturing and to design appropriate sealing systems for a nuclear waste repository.

In this paper, a model of tensile fracture is developed for the interpretation of microseismic events around underground openings and it is used to re-interpret the microseismic data from the Mine-by Experiment. It is shown that source dimension obtained with the proposed method is more realistic when compared to the available physical evidence. In this tensile model several key parameters are accounted for: the observed fractures grow parallel to the maximum stress direction, the fracture surface energy is not ignored, and the confining stress normal to the fracture is taken into account.

Backgound and Field Observations

Brittle fracture in rock masses has been investigated extensively both in the laboratory and in the field since the pioneering work of GRIFFITH (1921). The early laboratory work by BRACE et al. (1966), and BIENIAWSKI (1967) clearly showed that the brittle fracture on uniaxial compression test initiates very early in the loading process at stress levels of approximately 0.3 to 0.5 of the peak uniaxial load. Fracture initiation is defined as the process by which pre-existing cracks in a material start to grow or new cracks form. Microcracks commonly exist in rocks and fractures or joints exist in rock masses. In intact hard rocks, the sizes of these microcracks are often at the scale of the mineral grains. As pointed out by LAJTAI et al. (1990) and MARTIN and CHANDLER (1994), the initiation of fracturing in laboratory tests is not caused by shear slip as only dilation of the cylindrical samples is recorded with no axial shortening (Fig. 2b). This notion was confirmed by LOCKNER et al. (1992) who demonstrated using acoustic emission techniques that shearing of laboratory samples did not initiate until stress levels of approximately 0.7 of the peak strength was reached.

There are abundant experimental investigations which observe mode-I extension cracks forming and growing parallel or sub-parallel to the direction of the maxi-

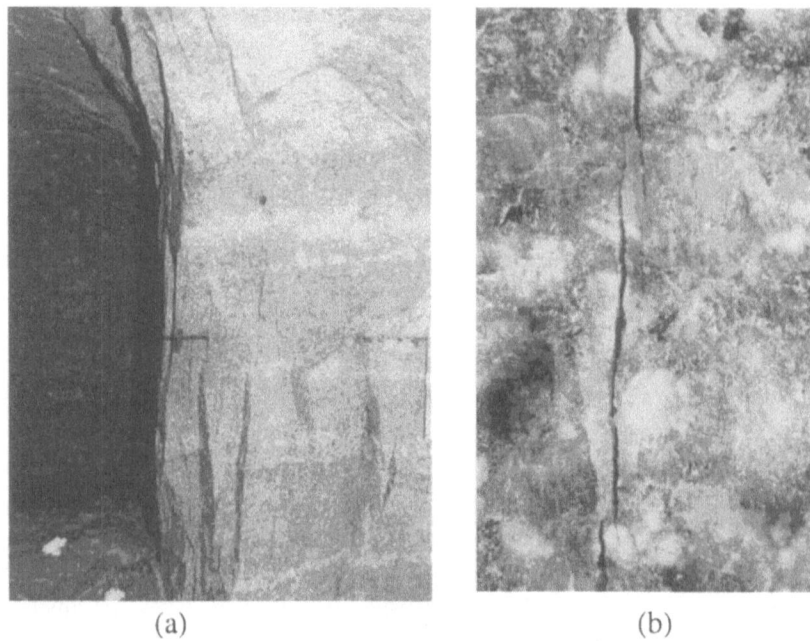

(a)   (b)

Figure 2
Example of observed fracturing in rocks: (a) Initiation of stress-induced fracturing (parallel to the tunnel
wall) observed around underground openings in a high stress environment. (b) Fractures observed in
samples of Lac du Bonnet granite loaded in compression. The direction of loading is in the direction of
the cracks. The grain size is approximately 2 to 5 mm and the cracks are approximately 5 to 10 mm long.

mum compressive stress under unconfined or confined conditions (e.g., WONG,
1982; HORII and NEMAT-NASSER, 1985; MYER *et al.*, 1992). Physical modeling of
brittle failure around circular openings has also been investigated extensively since
the early work of HOEK (1965). From tests of cylindrical openings in rock, EWY
and COOK (1990a,b) found that the fundamental fracture mechanism in the hollow
cylinders is the formation and growth of small extension cracks oriented parallel to
the tangential stress. LEE and HAIMSON (1993) also carried out testing of cubes of
Lac du Bonnet granite containing a borehole, and concluded, based on detailed
microscopic observations, that extension cracking is the dominant mode of failure
around the borehole in the region of maximum deviatoric stresses. These observations, in physical models using blocks of rock containing a circular hole, have also
been reported by MYER *et al.* (1992). LAJTAI *et al.* (1990) proposed that in order
to accurately model the initiation and propagation of fractures around circular
openings in physical models, the fractures had to be simulated with finite width
cracks. This allowed the fractures to grow in the direction of the maximum stress.
While many researchers have used the sliding crack model as the crack opening

force to develop the extension crack, LAJTAI *et al.* (1990) argued that there is no physical evidence for the notion that slip occurs prior to extension cracking. The crack opening force extends the crack in the direction of a predominant major principal stress ($\sigma_1 \gg \sigma_3$).

The mechanism governing the extension crack formation and growth is versatile. It may be driven by stress concentrations around inclusions, elastic mismatch between grains, contact between grains, or sliding along pre-existing cracks, fractures or joints. Recent advances in numerical modeling using particles to represent mineral grains have provided additional insight into brittle failure (CUNDALL *et al.*, 1996). The rock is treated as an assembly of grains bonded together at contact points, with the contact point acting like a pair of elastic springs which allow normal and shear relative motion. When either a tensile or shear-force limit is reached, the bonds break and can carry no tension thereafter. CUNDALL *et al.* (1996) carried out extensive numerical modeling of the Lac du Bonnet granite and showed that the non-uniform distribution of the grain size alone was sufficient to develop tensile failure internally in a sample loaded in deviatoric compression. By tracking the development of the fractures they were able to demonstrate that the dominant mode of cracking was caused by tensile bond failure. More recent work by DIEDERICHS (1998) using numerical models indicated that in compressional loading tensile cracking dominated shear cracking by a ratio of approximately 50:1. These models clearly established that it is the interaction of mineral grains in rock that is key to generating the tensile stresses in compressive loading and that the tensile cracking occurs in the region subjected to deviatoric compressive stresses even if confined by an overall confining stress $\sigma_3 \ll \sigma_1$. It is concluded that brittle rock failure in laboratory compressive tests is initiated by tensile cracking and as the number of tensile cracks increases, strong crack interaction among these tensile cracks will result in macro-scale shear failure of the rock.

Brittle fractures observed around underground openings typically occur as discrete fractures parallel to the boundary of the opening (Fig. 2a). Visual observations of fractures around underground openings in brittle rocks suggest that most of the fractures form by extension (tensile) processes (ORTLEPP, 1997). Early attempts to predict the extent of brittle failure around underground openings, using conventional frictional based failure criterion, met with limited success (MARTIN *et al.*, 1998). STACEY (1981), with extensive experience from South African mines, proposed an extension strain criterion for predicting the extent of fracturing of brittle rocks near underground openings. More recently MARTIN *et al.* (1998), using case histories for a wide range of rock mass conditions, showed that a criterion based on the initiation of extension fracturing, i.e., a critical deviatoric stress, provided a good estimate of the depth of brittle fracturing. The fracturing observed around the Mine-by test tunnel also supports this notion (MARTIN *et al.*, 1997).

Contrary to natural earthquakes in which the rock mass failure is mainly attributed to slip along pre-existing faults, the rock mass fracture process near underground openings is at least in part governed by extension cracking. Therefore, the focal mechanism related to microseismic events should differ from that of natural earthquakes. This paper builds on the extensive laboratory and field evidence, suggesting that brittle fracture near underground excavations is initiated and dominated by tensile fracture. Based on this evidence, a tensile model for the interpretation of microseismic events near excavations is presented. The authors realize that in reality many events are produced by combined modes of failure. However, for simplicity and to introduce the concept of a tensile failure model in an overall compressive stress field, it is assumed that all events fail in pure tension.

Interpretation of Microseismic Data

The interpretation of the small-scale microseismic events, such as those recorded during the excavation of AECL's 46-m-long Mine-by test tunnel, is usually based on earthquake seismology (e.g., BRUNE, 1970; RANDALL, 1973; MADARIAGA, 1976), assuming that the fractures form by shear slip (e.g., MCGARR, 1984; GIBOWICZ et al., 1991). FEIGNIER and YOUNG (1992, 1993), using moment tensor analysis, revealed that about 50% of the events recorded in the Mine-by experiment had significant non-shear components. Studies by URBANCIC et al. (1992) at Strathcona mine also showed that microseismic events near an advancing mining face were depleted of S-wave energy components, indicating that tensile fracture mechanisms are involved in rock failure in deep underground mines. GIBOWICZ et al. (1991), after analyzing the source parameters of microseismic events from the shaft at the URL, explained that in the case of the local granite, the fractures might be first initiated by a mechanism rich in tensile components prior to shear failure along the same fractures. The important point is that the practical observations of extension fracturing (both in laboratory and *in situ*) are not compatible with the traditional shear models that are commonly used to interpret microseismic events. If progress is to be made in relating source dimensions from microseismicity and observed rock mass fracturing, this incompatibility must be resolved.

Fault Slip Model

All seismic events, whether they are due to slip on a single fault or on various fractures, or are formed in tension, cause straining of a volume of rock surrounding the seismic source (even if the source is of a planar character). The volume of coseismic inelastic deformation can be estimated (MENDECKI, 1997) as

$$V = \frac{M_0}{\Delta\sigma} \tag{1}$$

where M_0 is the seismic moment and $\Delta\sigma$ is the stress drop. Alternatively, it can be defined by the apparent volume as

$$V_A = \frac{M_0}{2\Delta\sigma_A} = \frac{M_0^2}{2GE_0} \tag{2}$$

where $\Delta\sigma_A$ is the apparent stress drop, E_0 the seismic energy, and G the shear modulus of the rock mass near the source. The apparent volume, which is a scalar quantity, is also an indicator of the volume of rock affected by inelastic deformation due to the seismic event. The radius of a sphere containing this volume of inelastic deformation can be found from

$$r = \sqrt[3]{3V_A/(4\pi)}. \tag{3}$$

This concept implies that the amount of inelastic shearing is largest near the center and reduces to negligible values at the edge of the sphere. The size of the fracture which causes the inelastically deformed volume of rock must be less than the radial extent of the fracture. The radius of a sphere exceeding a certain threshold of inelastic strain will depend on the chosen threshold value but is less than the radius defined by equation (3)

$$r_{\text{threshold}} = \eta r$$

or

$$V_{\text{threshold}} = \eta^3 V_A \tag{4}$$

where η is a coefficient which depends on the threshold value. In this manner, the source size is not defined as the size of a physical fracture but as the consequence of fracturing which causes the inelastic strain.

Source Dimension

One of many possible models is to take the radii from equation (3) as the source dimensions. This is, however, not correct as the source size must be considerably smaller than the size of the rock mass which experiences inelastic straining. Alternatively, by application of a kinetic or quasi-dynamic model, the radius r_0 of a circular fault could be used. It is inversely proportional to the corner frequency f_c of S wave (GIBOWICZ and KIJKO, 1994)

$$r_0 = \frac{K_c\beta_0}{2\pi f_c} \tag{5}$$

where K_c is a constant which depends on the source model and β_0 is the S-wave velocity in the source area. Equation (5) applies to the tensile model of SATO (1978) and the shear models of BRUNE (1970) and MADARIAGA (1976), only in that the value of the coefficient K_c is different for different models. For BRUNE's (1970) source model, only S waves are considered and the constant K_c is set equal to 2.34, a constant that is independent of the angle of observation. In MADARIAGA's (1976) model, the coefficient K_c is a function of the angle of observation, i.e., the angle between the normal to the fault and the takeoff direction of P or S waves. If the average value of the coefficient $K_c = 1.32$ is used, the source dimension by Madariaga's model leads to 56% smaller sources. It has been reported that Madariaga's model seems to provide more reasonable estimates of source dimensions than the Brune model when compared to physical observation in some mines and underground openings (GIBOWICZ and KIJKO, 1994; TRIFU *et al.*, 1995). In these shear models, the fracture surface energy is neglected, which is a reasonable assumption for large-scale seismic events such as earthquakes. As will be illustrated later, for small-scale microseismic events, however, the fracture surface energy cannot be ignored for the determination of the source dimension because it constitutes a dominant component of the energy balance equation.

Interpretation of Microseismic Events at URL Based on Shear Model

Over 25,000 events were detected and some 3500 events (or 14%) were source located at URL test tunnel (READ and MARTIN, 1996). To estimate the rock mass damage from the microseismic measurements, a 3.5-m-thick slice (approximately one tunnel diameter) was chosen for analysis between chainage 21.03 m and 24.59 m along the tunnel axis. 804 seismic events are located in this region. If we assume that the events are uniformly distributed along the tunnel axis and the overall percentage of 14% located events is generally applicable, then about 4500 to 5000 additional events must have occurred in this region. Figure 3 shows the cross-section location of these 804 microseismic events and the geometry of the notch which formed as the tunnel was advanced (refer to Fig. 1 as well). The deviatoric stress contours ($\sigma_1 - \sigma_3$) from a 2-D linear elastic analysis are also shown in this figure. It is seen that most microseismic activity in the roof of the tunnel is concentrated in the region where the deviatoric stress exceeds the crack initiation threshold of about 70 MPa (or $(0.4 \pm 0.1)\sigma_c$) and where the notch eventually forms.

The physical sizes of the fractures associated with the microseismic events were first estimated from the traditional shear models similar to the study by GIBOWICZ *et al.* (1991) for data from the URL shaft. The radii given from equation (3) vary from 0.172 m to 1.833 m. The radii given from equation (5) of Brune's model vary from 0.21 m to 2.69 m and the diameter distribution is shown to scale in Figure 4a. If Madariaga's model is used, the source radii will be about half smaller than those from Brune's model (Fig. 4b). Of course, these fractures would not form in the cross-sectional plane but rather in a plane perpendicular to the cross-section plane and at angles to the excavation boundary as shown for one source at point A in Figure 4a. It can be seen from the figures that, whether the estimated crack size is obtained from the Brune's model or from Madariaga's model, the source dimensions are unrealistically large, with many large deep-seated events in areas of scant visual damage (e.g., in the floor in Fig. 1). As was explained earlier, one may view the apparent volume as an indicator of the volume of rock experiencing inelastic deformation and thus argue that the physical source is substantially smaller. A correction factor of 1/10 or less would be required to approach meaningful dimensions. Unfortunately, there is no basis for the selection of such a correction factor. Brune's model gives fracture sizes of comparable size and thus also provides unrealistic fracture sizes. This is consistent with the views expressed by GIBOWICZ and KIJKO (1994) and REVALOR et al. (1990) which explain that source sizes given from the Brune's source model are unrealistically large for small microseismic events. Although they indicate that Madariaga's model is fine for their observation, this cannot be upheld in the case of Mine-by Experiment where detailed investigations did not locate any fractures (other than grain size) in the region of the source located microseismic events (Fig. 1).

As was discussed above, the physical fracture sizes obtained from the traditional models are not in agreement with site observation. This suggests that there are limitations to these source models. Cracking of brittle rock on compression occurs by stable tensile cracking and this process is a stress-controlled phenomena. For example, the presence of confining stress tends to stabilize the crack growth, i.e., the lengths of the tensile cracks depend on the confining stress (HOEK, 1968). Details of the effect of confining stress on the fracture size will be discussed later in this paper.

Another limitation of the existing source models is that shear slip is assumed to drive the failure process, which may be valid for natural earthquakes and large fault

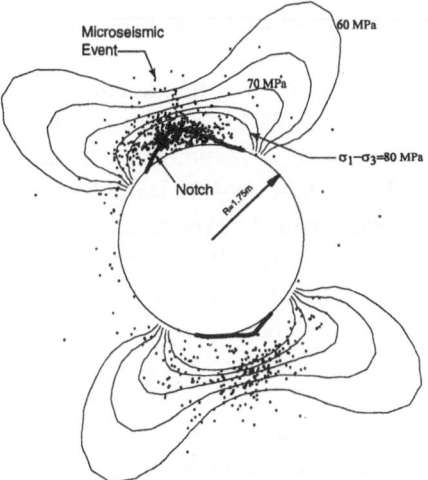

Figure 3
Cross-section location of the microseismic events and fracture limit defined by the constant-deviatoric stress criterion.

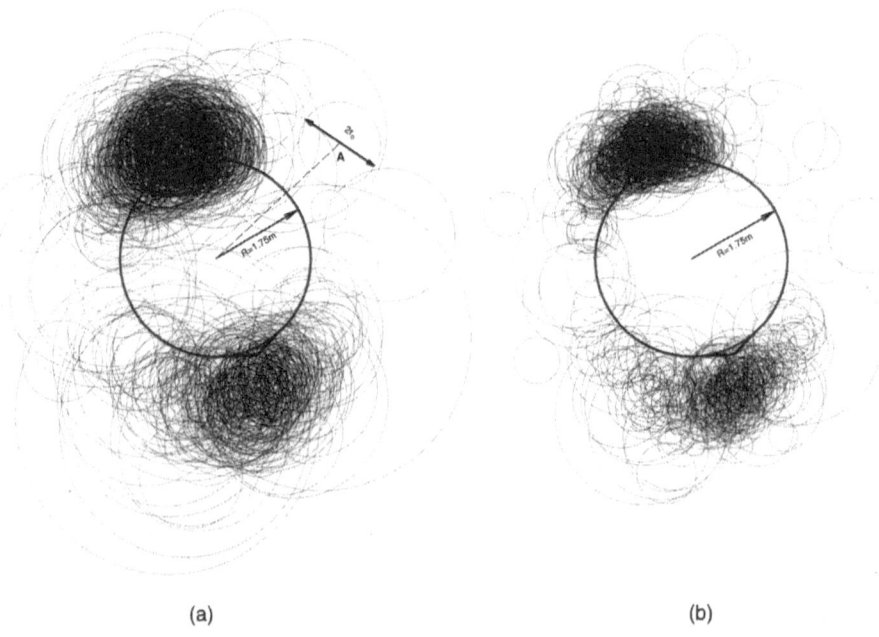

Figure 4
Crack size distribution from shear models: (a) Brune's model; (b) Madariaga's model. The latter is obtained based on a factor of 0.56 of the first one.

slip type tremors in mines. There is, however, growing evidence of observed energy depletion in S waves recorded in mines, implying that tensile failures, or at least shear failures with a strong tensile component, are frequently generated near underground excavations. S-wave energy E_s versus P-wave energy E_p is presented in Figure 5 for URL data. The frequency of E_s/E_p ratio is also shown in Figure 5 as an insert. As can be seen, the ratio of the radiated seismic energy components E_s and E_p ranges from 1 to 80, with 78% of the 804 events registering ratios below 10. Note that the percentage of the E_s/E_p ratio, which is smaller than 10 for the seismic events recorded at the URL shaft, is 40% (GIBOWICZ *et al.*, 1991). The ratio of E_s/E_p is an important indicator of the type of focal mechanism responsible for the generation of seismic events. SATO (1978) has shown that the P- and S-wave energies are approximately equal for his tensile failure model. In pure shear, the S-wave energy is considerably larger than the P-wave energy. Hence, it is commonly accepted that if $E_s/E_p < 10$, the cracking process involves a tensile failure component (GIBOWICZ *et al.*, 1991; GIBOWICZ and KIJKO, 1994). If $E_s/E_p > 20$ to 30, the energy radiated in P waves is only a small fraction of that in S waves, and shear failure dominates this type of failure process (BOATWRIGHT and FLETCHER, 1984). Many mining-induced seismic events that have large moment magnitudes also have high E_s/E_p ratios and thus can be analyzed by shear models (MCGARR, 1984; GIBOWICZ and KIJKO, 1994).

Figure 6 presents the relationship between E_s/E_p ratio and the location of each event relative to the tunnel face position X normalized with respect to the tunnel diameter $2R$. It can be seen that most of the microseismic activity occurs near the advancing tunnel face and decreases rapidly with distance from face to about $X/2R = -2$. This is the area where the stress changes from a 3-D to a 2-D state. Most events with $E_s/E_p > 30$ concentrate around the site for $X/2R = -1$, and events exceeding a ratio of 20 are predominately observed in the region between $X/2R = 0$ and $X/2R = -1$. The slabbing associated with notch formation is observed to start at $X/2R = -0.14$ to $X/2R = -0.28$ (MARTIN et al., 1997).

It can be seen from the above arguments that the cracking mechanism is rich in tension at the Mine-by test tunnel, and thus a tensile crack model might provide a better estimate of fracture size.

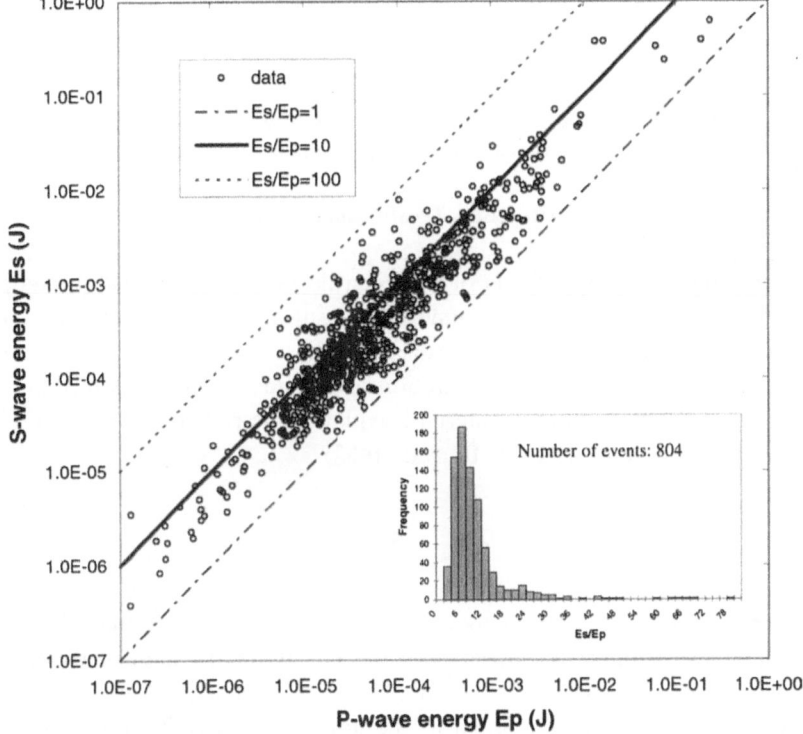

Figure 5

S-wave energy versus P-wave energy ratio at the URL between chainage 21.03 m and 24.59 m. The frequency distribution of E_s/E_p is also shown as an insert.

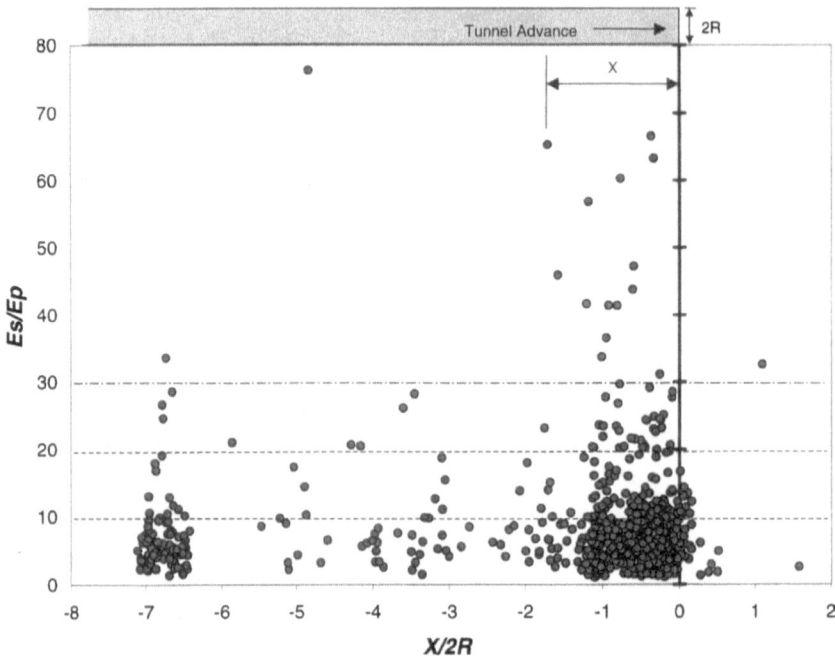

Figure 6
Distribution of E_s/E_p ratio at the URL. The location of each event relative to the tunnel face position X is normalized with respect to the tunnel diameter $2R$.

Proposed Tensile Source Model

According to the hypothesis of pure tensile failure in an overall compression stress field introduced earlier, assume that a slit-like crack (fracture) is suddenly introduced in an infinite uncracked rock. The elastic strain energy change per unit thickness caused by the sudden generation of the crack in the rock medium and the relaxation of the surrounding material is equal to the potential energy and is given by (e.g., GRIFFITH, 1921, 1924; BROEK, 1982)

$$U_c = -\frac{(1-v^2)\pi\sigma_n^2 a^2}{E} \tag{6}$$

where σ_n is the stress normal to the crack, E and v are Young's modulus and Poisson's ratio of the rock respectively, and a is the half length of the crack. This can also be explained differently. Before the crack is generated, we can view it as a potential crack with traction equal to the normal stress acting on the crack edge. Once the crack is generated by tensile fracturing (the mechanism of tensile stress build-up is versatile and will not be addressed here), the forces normal to the crack edges eventually become zero. It is evident that work will be released as radiated energy upon the sudden generation of the crack. Ignoring the mechanism that generates the tensile crack, the stress term that comes into Equation (6) should be σ_n.

Energy is absorbed during the fracture process because a change in the elastic surface energy is required to overcome the cohesive molecular forces at the crack tips. This surface energy is given by

$$U_s = 4a\gamma_s \tag{7}$$

where γ_s is the specific surface energy.

If we assume that the observed or measured seismic energy is the energy imbalance (kinetic energy), i.e., the energy difference between the potential energy and surface energy when a crack is generated, then we derive

$$-\frac{(1 - v^2)\pi\sigma_n^2 a^2}{E} + 4a\gamma_s + E_0 = 0 \tag{8}$$

where E_0 is the energy imbalance, assumed to correspond to the recorded seismic energy E_s^*.

In order to estimate the tensile source based on this model and the microseismic monitoring results, the following assumptions are made:
- Each microseismic event corresponds to a single Type-I planar crack or fracture. This Type-I crack is planar, has no dislocation, and is parallel to the maximum principal stress direction (σ_p) or perpendicular to the normal stress ($\sigma_n < \sigma_p$).
- The seismic energy is equal to the kinetic energy generated in the formation of the crack. This assumption can be relaxed if the energy loss due to wave attenuation is to be considered. Generally, we can express $E_s^* = \chi E_0$, where χ is a factor which takes into account the energy loss due to propagation.
- The initial crack length is small and can be neglected (the whole crack is suddenly generated) and energy barriers exist so that cracks grow to a finite size.

In reality, the cracks related to microseismic events are three-dimensional. In a three-dimensional setting with penny-shaped cracks, the crack radius a can be calculated by solving a cubic equation:

$$-\frac{8(1 - v^2)\sigma_n^2}{3E} a^3 + 2\pi\gamma_s a^2 + E_0 = 0. \tag{9}$$

It is interesting to note that the crack length is proportional to the seismic energy in the form of $a^3 \propto E_0$ (Equation 9), if the surface energy is neglected (a common assumption made in earthquake modeling (SCHOLZ, 1990)). Most importantly, contrary to traditional source models, the influence of normal stress σ_n at the source location is explicitly included in this model.

As had been stated by GRIFFITH (1921), it is possible that all or some of the elastic potential energy is balanced by the crack resistance or surface energy in such a manner that a crack grows slowly in a silent fashion. By definition, this is the case for a crack of critical length (SCHOLZ, 1990) where all energy is consumed internally and no energy is emitted. Consequently, the formation of cracks of critical length cannot be recorded by microseismic systems because $E_0 = 0$, hence $E_s^* = 0$.

The determination of the normal stress σ_n at the source location deserves special consideration. It is the minimum stress at the time and point where the seismic event occurs. In reality, as a tunnel or mining face advances, the loading path of a point in space is very complex, involving stress increase, decrease and stress rotation close to the advancing excavation (MARTIN, 1997). If we accept the crack initiation criteria $\sigma_1 - \sigma_3 \cong (0.4 \pm 0.1)\sigma_c$ introduced earlier for use at the URL (MARTIN and READ, 1996), three-dimensional stress analysis can be performed to determine the normal stress σ_n at which cracking starts. Alternatively, the minimum stress can be obtained from numerical models at the time of the recorded seismic event.

Knowing the rock properties of E, v, and K_{IC}, we can estimate $2\gamma_s$ from $((1 - v^2)/E)K_{IC}^2$ (BROEK, 1982). Assume the rock properties of $E = 65$ GPa, $v = 0.25$, the predicted effect of σ_n and K_{IC} on the crack radius is presented in Figure 7. At lower confinement levels cracks grow longer. This is consistent with laboratory experiment results (e.g., HOEK, 1968). Another important observation from this parameter study is that for a given set of stress and material property, there is a critical crack length. In other words, there is no seismic energy radiation if the actual crack length is equal to the critical crack length. For example, for $\sigma_n = 11$ MPa and $K_{IC} = 2.5$ MPa$\sqrt{m}$, cracking with radius equal to 0.061 m will not emit

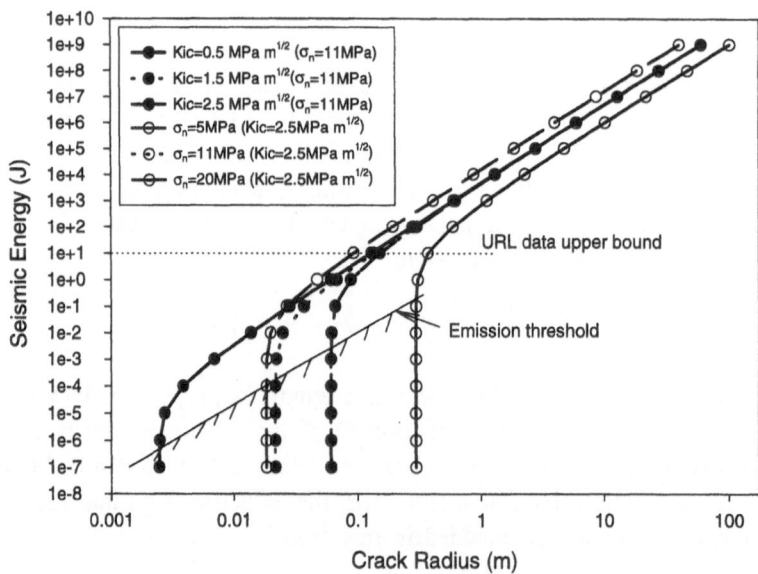

Figure 7
Influence of K_{IC} and σ_n on the crack radius and seismic energy relationships ($E = 65$ GPa, $v = 0.25$).

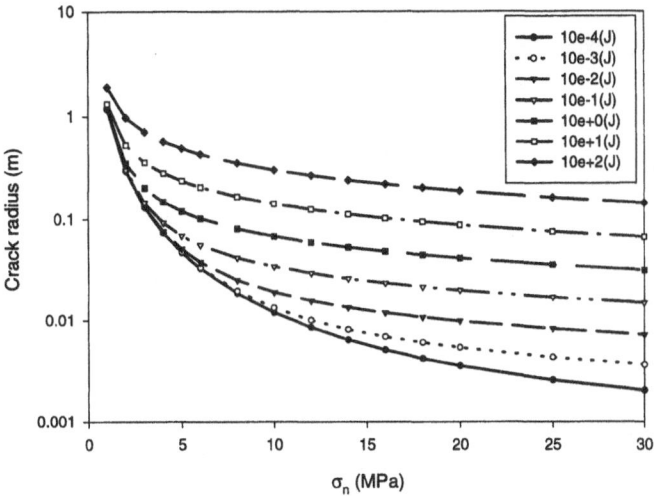

Figure 8
Effect of confining stress on the crack radius.

any seismic energy. In this case, the elastic potential energy is balanced by the crack resistance energy in such a manner that cracks grow in a silent fashion. Therefore, an emission threshold exists as shown in Figure 7. This threshold is inclined, indicating that only relatively long cracks can be detected near excavations at locations with low confinements. Only at energy levels of about one order of magnitude higher energies than this threshold, is the logarithm of crack length proportional to the energy imbalance E_0.

Whereas the Young's modulus of the host rock has little effect on the crack radius, it is shown here that the specific energy γ_s, or the fracture toughness K_{IC} has a strong influence on the crack length when the seismic energy is small. It has little to no effect when the seismic energy is large (Fig. 7). This confirms that the surface energy can be neglected for natural earthquakes or large mining-induced events (with seismic energy exceeding 1 to 10 J for the selected rock properties) but not for the interpretation of microseismic events with small seismic energy releases.

Figure 8, derived from Equation (9) with $K_{IC} = 1.0$ MPa$\sqrt{m}$, shows the dependence of the crack radius on the normal stress σ_n for different energy imbalance values ranging over six orders of magnitude. Again, the crack size and normal stress relationship is consistent with experimental results (HOEK, 1968). When the confining stress is small, e.g., near the excavation wall, the source size does not depend on the energy imbalance. All source sizes are large and thus depend insignificantly on the recorded seismic energy. For example, if $\sigma_n < 2$ MPa, the crack radii range from 0.5 to 1.0 m for emitted energies between 10^{-4} J and 10^{+2} J. This implies that the confining pressure dominates the size and thus the degree of damage near excavation walls. Unless the confining stress is accurately known at an event location, this method cannot be used to quantify damage near the surface of an excavation. However, inside the rock mass, e.g., at locations with $\sigma_n \geq 5$ MPa, rock damage is sensitive to the released energy and damage should be quantifiable if seismic energy measurements are combined with stress calculations. This is illustrated by the following case study.

Application of Tensile Source Model

Source Dimension at the URL Test Tunnel

The radiated seismic energy at URL ranges from 10^{-7} to 10^1 J (MARTIN and READ, 1996). The upper bound is shown in Figure 7. It is seen that most of these events fall near the predicted emission threshold. Hence, many of these small microseismic events come from cracks that are near the critical crack length. As was stated earlier, in this range, the crack length is dominated by K_{IC} and very sensitive to σ_n.

A 3-D elastic model using Map3D (WILES, 1996) was constructed of the cylindrical tunnel and used to determine the normal stress σ_n at the seismic event locations when a crack is created or initiated. There are eleven excavation steps in the model, and the minimum stress σ_n corresponding to the point at which the crack initiation threshold $\sigma_1 - \sigma_3 \geq 70$ MPa is reached, was picked and used to calculate the source size. The initial *in situ* stress state was $\sigma_1 = 60$ MPa, $\sigma_2 = 45$ MPa, $\sigma_3 = 11$ MPa with σ_2 parallel to the tunnel axis and σ_1 rotated by 15 degrees anti-clockwise from horizontal (MARTIN and READ, 1996; MARTIN, 1997). The source sizes calculated using $E = 65$ GPa, $v = 0.25$, $K_{IC} = 0.5$ MPa m$^{1/2}$, are given in Figure 9. The crack radii estimated in this manner from the measured energy, range from 0.00074 to 0.556 m. In Figure 9, the largest circle near the roof surface corresponds to a 1.112-m long crack. Again, these cracks are not oriented in the cross-sectional plane but, according to the model, are parallel to the major principal stresses at the event locations.

It can be seen from Figure 9 that most cracks in the roof are concentrated near the wall where the notch eventually formed. Further away from the tunnel surface, the normal stress gradually increases and the microseismic damage both in size and density is considerably smaller. This agrees well with the observed rock mass behavior reported by MARTIN and READ (1996), both in terms of notch formation (shown in Fig. 9) and as indicated from extensometers, exhibiting inelastic deformation near the excavation surface. Tests on hollow cylinder (25.4 mm in diameter) samples under non-uniform stress state by EWY and COOK (1990a,b) also show that the density of cracking decreases radially away from the boundary, again reflecting the effect of increasing confining (normal) stress.

One might argue that the source radii shown in Figure 9, in general, are too small because energy loss, as explained earlier, is not considered. Visual field observation from boreholes or a slot excavation show no sign of extensive crack damage in the rock mass except very close to the notch (see Fig. 1). This confirms that the rock mass outside the notch must be undamaged. This is consistent with Figure 9, even if it is considered that actually 7.1 times more than analyzed events must have occurred in this volume (because only 14% of recorded events are source located). There are four large-sized crack events displayed in Figure 9, all of them are attributed to the fact that the stresses perpendicular to the crack surfaces are very small at those locations. If the cracking is near the tunnel surface, this would lead to spalling by crack coalescence, explaining the slabbing observed in the roof notch. As explained earlier, crack growth is largest at low σ_n. There are relatively fewer near-surface events in the floor and this is reflected in the non-symmetric notch formation. The large predicted events inside the rock mass are unrealistic

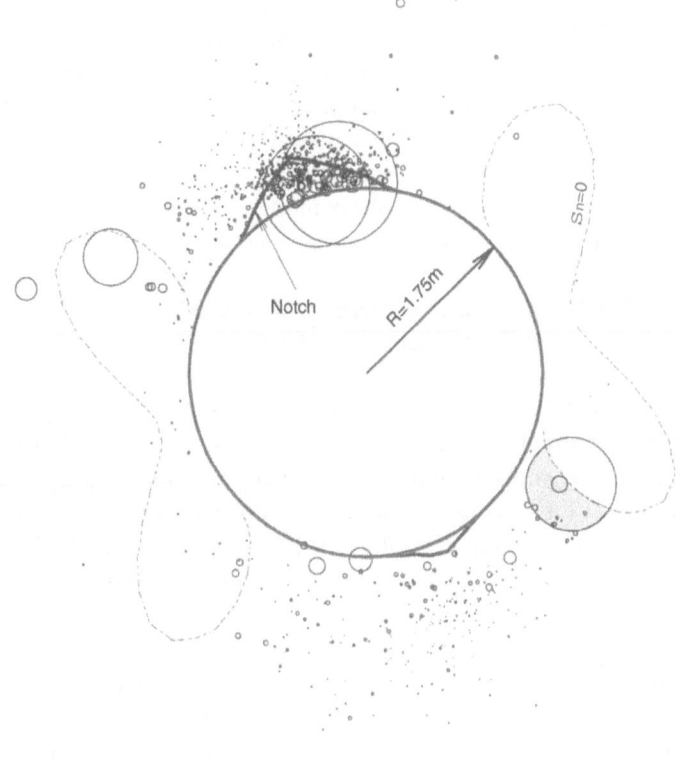

Figure 9
Crack size distribution from the tensile model ($\chi = 1$). The length of a crack is indicated by the diameter of the circle but the crack is oriented parallel to the maximum stress direction.

because the model does not take stress gradients in σ_n into account. For example, the large source size for the event at the lower right of the tunnel is attributed to the fact that the event center is very close to the zero stress contour line (shown in Fig. 9 as a dashed line) predicted by the elastic continuum model without tension cutoff. In reality, this minimum stress changes its direction and magnitude in space, and the propagation of a crack will stop to grow as soon as the minimum stress becomes larger in the shaded area. It should be noted that the same shortcoming applies to other source mechanism models because most of the source dimensions are too large to meet the implied uniform stress assumption (see Fig. 4). Bearing in mind these shortcomings, these "unrealistically large" events should be discounted, or an average σ_n over the source size should be used to calculate the source size in an iterative manner.

Should seismic loss be considered, then the correction factor χ is less than 1. Here, χ is defined as the ratio of the recorded seismic energy over the radiated seismic energy. It is difficult to determine the exact number of χ, but, as an extreme case, the crack radii are re-calculated with a correction factor of $\chi = 0.1$ and their distribution is shown in Figure 10 (compare with Fig. 9 for $\chi = 1$). These crack sizes are larger, particularly in areas of low confinement (refer to the discussion of Fig. 8) but still reasonably sized as there is a good visual agreement of the event clustering in the notch region. The overlapping events accurately define the severely damaged region in and near the notch. It is seen that the actual crack sizes should be between the crack sizes shown in Figures 9 and 10. The source radii, shown in Figure 4a, were estimated by Brune's shear model assuming $\chi = 1$. If the χ is less, even bigger and more unrealistic source sizes would be predicted by the shear models.

The relationship between seismic energy and source radius is shown in Figure 11, for shear (Brune) and for the tensile model (Eqn. 9). The data for the tensile model does fit the extrapolation of the empirical relationship obtained for large rock bursts (TALEBI, 1993; KAISER and MALONEY, 1997). This is largely attributed to the fact that the normal stress differs in the tensile model for each event location although not for the conventional shear model. The relationship between the calculated normal stress and source radius, for three ranges of seismic energy, is presented in Figure 12, for the case of $\chi = 0.1$ (compare to Fig. 8). The measurements are compared to predicted relationships for two extreme energy levels, confirming the earlier finding that, even if the energy is small, the corresponding crack size can be relatively large at low confining stress. If the influence of normal stress variation were not considered, the resulting trend of energy-source radius relation for tensile model would be the same as that of the shear model (see Figs. 7 and 11). Figure 13 presents the correlation between the source radii for the tensile model ($\chi = 1$ and 0.1) and Brune's shear model. The trend lines are drawn according to the ratios of the average radii from the shear model and tensile model. It is seen that the source sizes for the shear model are at least 29 times larger than those of the tensile model (Fig. 13b).

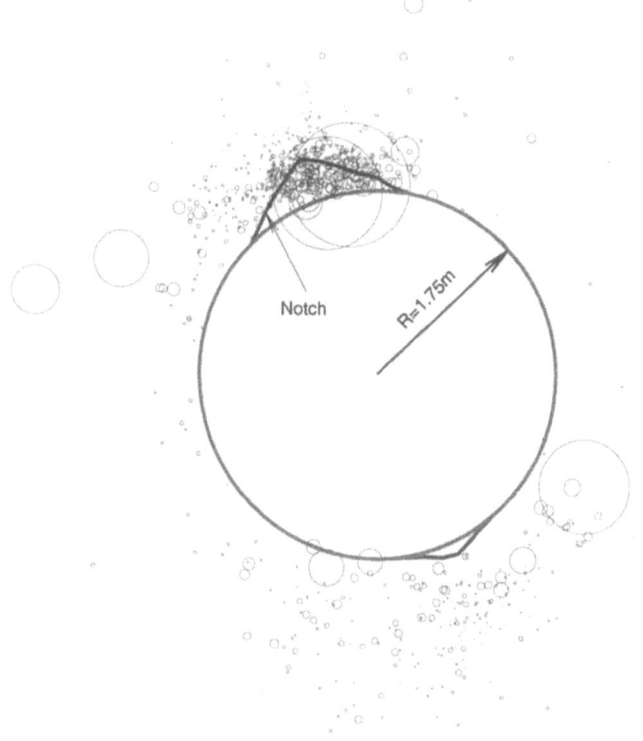

Figure 10
Crack size distribution from the tensile model ($\chi = 0.1$).

Figure 14 presents the source dimension ($\chi = 0.1$) distribution for four different ranges of E_s/E_p ratio for all recorded seismic events in the investigation volume (shaded area). Events with $E_s/E_p = 1$ to 10 locate inside and just outside the final notch boundary in the roof (Figs. 14a and b), and show relatively uniform source location distribution. The majority of the events with $E_s/E_p > 20$ are found outside or at the edge of the final notch boundary (Fig. 14d). If we plot the events distribution in two stages of excavation periods (Fig. 15) before and after point B (face chainage 24.48 m), the tunnel face where Round 20 excavation ended and the notch formation was visually observed, we find that as tunnel face advanced from B to C, most of the events with $E_s/E_p = 10$ to 20 are located outside the final notch region (Fig. 15c). All of the events with $E_s/E_p > 20$ are located outside the final notch boundary (Fig. 15d). It can be seen that seismic events with large S-wave energy components are generated at the interface between the rock heavily dam-

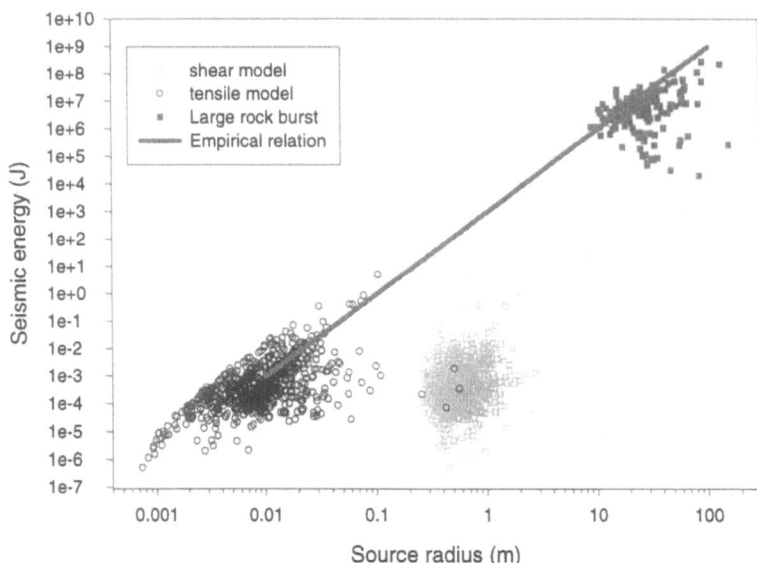

Figure 11
Source radius and seismic energy relations for the tensile model ($\chi = 1$), shear model at the URL and some rockburst data.

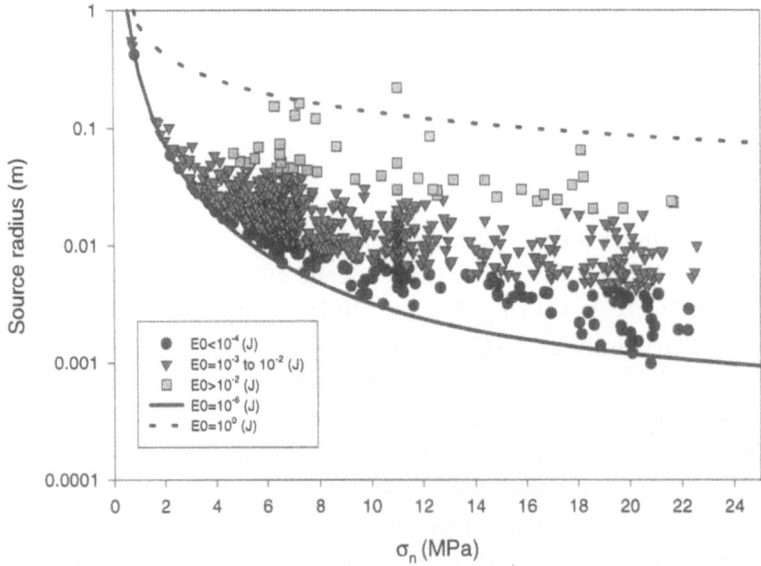

Figure 12
Effect of confining stress on the source radius at the URL ($\chi = 0.1$).

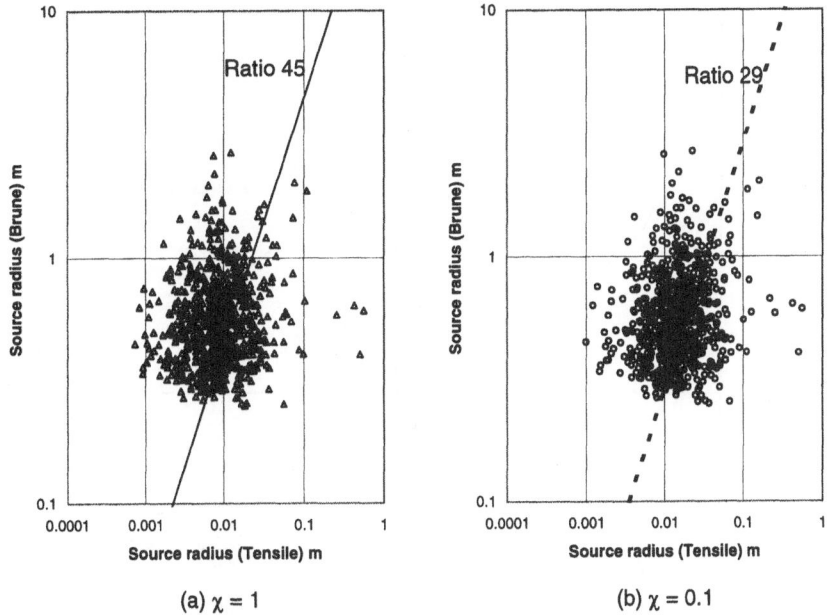

Figure 13
Correlation between the source radii from the tensile model and Brune's shear model: (a) $\chi = 1$; (b) $\chi = 0.1$.

aged by tensile cracking and the relatively undamaged rock. The tensile events seem to precondition the ground leading to stress relief in this zone and stress concentration at its edge. At the notch boundary stresses are thus sufficient to cause shear failure. It seems to follow that the notch boundary can be defined as the interface between tensile and shear failure clustering. The spatial analysis of seismic source parameters indicates that the failure mechanisms varied with the relative position to the advancing excavation face. This is consistent with the observation from Strathcona mine (Sudbury, Ontario, Canada), where events with large non-shear components were observed close to the advancing mining face, and events with large shear components were located outlying from the mine face (URBANCIC et al., 1992).

The results of this study will open the way to determine rock-mass damage from microseismic observations. Knowing from visual observations that stress-induced cracks are parallel to the maximum stress direction or parallel to the excavation wall, the crack density tensor can be calculated, and the rock-mass damage can be quantified. By combining the effects of cracks on effective moduli and strength, and by continuously upgrading the information from microseismic measurements, it should be possible to analyze the nonlinear stress-strain response of rock masses in the field.

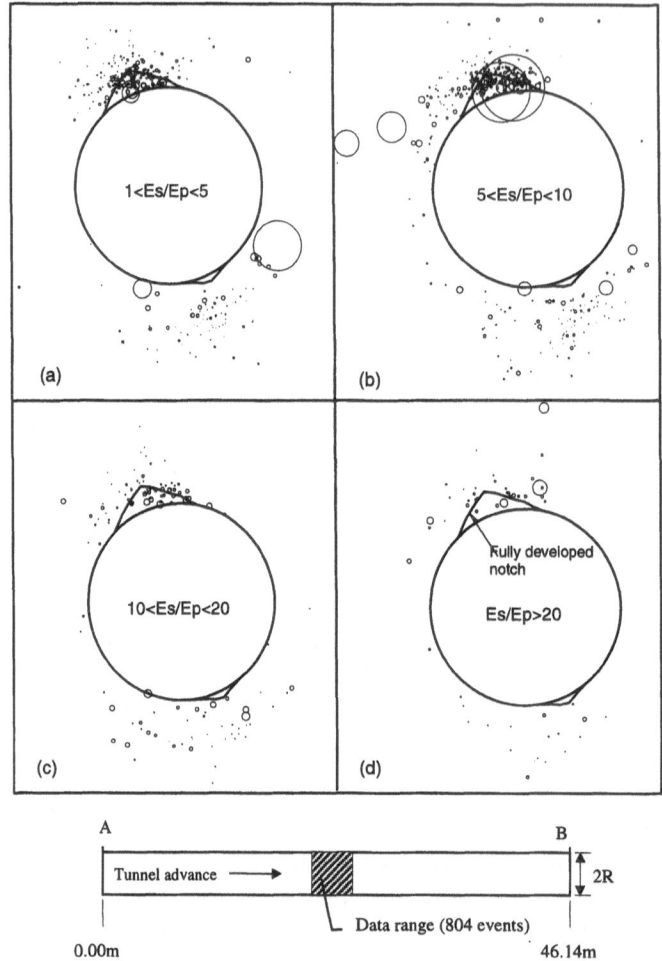

Figure 14
Distribution of source size ($\chi = 0.1$) in space with different E_s/E_p ratios: (a) $E_s/E_p = 1$ to 5; (b) $E_s/E_p = 5$ to 10; (c) $E_s/E_p = 10$ to 20; (d) $E_s/E_p > 20$.

Conclusions

Traditional source radius models, based on shear failure, provided unrealistically large source radii for the microseismic events recorded at the AECL's Mine-by Experiment, when compared to an observation of damage. A detailed analysis of 804 microseismic events recorded near the center of the experiment suggested a tensile cracking mechanism. A tensile model was developed that gave source radii consistent with field observations.

The model assumes that the measured seismic energy is a portion of the total radiated energy, which is equal to the difference between the elastic strain energy

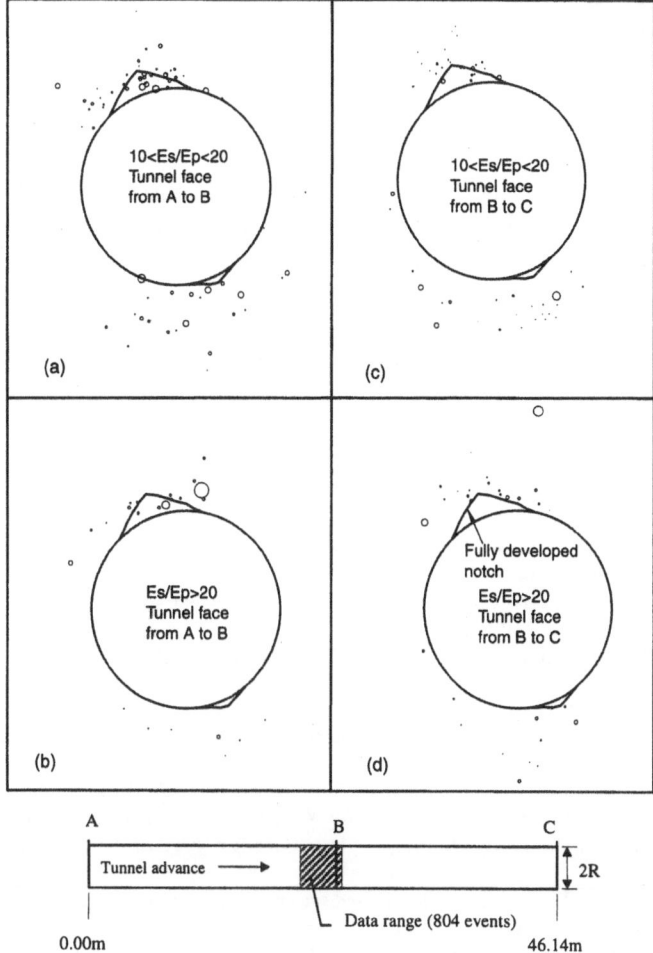

Figure 15

Distribution of source size ($\chi = 0.1$) in space with different E_s/E_p ratios and excavation windows: (a) $E_s/E_p = 10$ to 20 tunnel face advances from A to B; (b) $E_s/E_p > 20$ tunnel face advances from A to B; (c) $E_s/E_p = 10$ to 20 tunnel face advances from B to C; (d) $E_s/E_p > 20$ tunnel face advances from B to C.

change and the surface energy. The elastic strain energy is calculated from the model by putting a flat crack in an infinite solid under uniform far-field stress. The crack size estimated from this model depends on surface energy and on the normal stress at the point of crack formation, which led to more realistic sizes when compared to field observations, than those determined by traditional shear slip models.

It has been shown from the E_s/E_p ratio that not all seismic events occur in tension. For simplicity and to introduce the tensile model, all events at URL have been modeled as tensile failure in this study. Future direction could be the combination of tension and shear failure models to determine the source sizes of microseismic events.

Acknowledgments

This work has been supported by a strategic research grant of the Natural Science and Engineering Research Council (NSERC) of Canada on the Mitigation of Violent Failure Processes in Deep Excavations. The authors are grateful to Atomic Energy of Canada Limited (AECL) for providing detailed data for this investigation. The authors wish to thank Dr. D. Collins for providing the source-located microseismic data from the URL. Comments from an anonymous reviewer enhanced the quality of this paper.

REFERENCES

BIENIAWSKI, Z. T. (1967), *Mechanism of Brittle Fracture of Rock, Parts I, II and III*, Int. J. Rock Mech. Min. Sci. and Geomech. Abstr. *4* (4), 395–430.

BOATWRIGHT, J., and FLETCHER, J. B. (1984), *The Partition of Radiated Energy between P and S Waves*, Bull. Seismol. Soc. Am. *74*, 361–376.

BRACE, W. F., PAULDING, B., and SCHOLZ, C. (1966), *Dilatancy in the Frature of Crystalline Cracks*, J. Geophys. Res. *71*, 3939–3953.

BROEK, D., *Elementary Engineering Fracture Mechanics* (Martinus Nijhoff Publishers, the Hague 1982).

BRUNE, J. N. (1970), *Tectonic Stress and the Spectra of Seismic Shear Waves from Earthquakes*, J. Geophys. Res. *75*, 4997–5009.

CUNDALL, P. A., POTYONDY, D. O., and LEE, C. A., *Micromechanics-based models for fracture and breakout around the Mine-by tunnel*. In *Proc. Int. Conf. on Deep Geological Disposal of Radioactive Waste* (eds. Martino, J. B., and Martin, C. D.) (Canadian Nuclear Society, Toronto 1996) pp. 113–122.

DIEDERICHS, M. S. (1998), *Rock mass stability and support design for deep excavations in hard rocks* (in progress). Ph.D. Thesis, Dept. of Civil Eng. University of Waterloo, Waterloo, Ontario, Canada.

ECOBICHON, D., HUDYMA, M., and LAPLANTE, B. (1992), *Understanding Geomechanics Problems through Microseismic Monitoring at Lac Shortt*, Rock Mechanics and Strata Control Session, Proceedings of the 94th Annual General Meeting of the CIM, Montreal, April, pp. 1–13.

EWY, R. T., and COOK, N. G. W. (1990a), *Deformation and Fracture around Cylindrical Openings in Rock—I, Observations and Analysis of Deformations*, Int. J. Rock Mech. Min. Sci. and Geomech. Abstr. *27*, 387–407.

EWY, R. T., and COOK, N. G. W. (1990b), *Deformation and Fracture around Cylindrical Openings in Rock—II, Initiation, Growth and Interaction of Fractures*, Int. J. Rock Mech. Min. Sci. and Geomech. Abstr. *27*, 407–427.

FALLS, S. D., and YOUNG, R. P., *Comparison of excavation disturbance around deep tunnels in hard rock using acoustic emission and ultrasonic velocity methods*. In *Proc. Excavation Disturbed Zone Workshop* (eds. J. B. Marino, and Martin, C. D.) (Canadian Nuclear Society, Toronto 1996) pp. 77–86.

FEIGNIER, B., and YOUNG, R. P. (1992), *Moment Tensor Inversion of Induced Microseismic Events: Evidence of Non-shear Failures in the $-4 < M < -2$ Moment Magnitude Range*, Geophys. Res. Lett. *19* (14), 1503–1506.

FEIGNIER, B., and YOUNG, R. P. (1993), *Source Mechanism Studies at the Underground Research Laboratory*, Report to AECL, RP021AECL, Queen's University, Kingston, Ontario.

GIBOWICZ, S. J., and KIJKO, A., *An Introduction to Missing Seismology* (Academic Press, 1994).

GIBOWICZ, S. J., YOUNG, R. P., TALEBI, S., and RAWLENCE, D. J. (1991), *Source Parameters of Seismic Events at the Underground Research Laboratory in Manitoba, Canada: Scaling Relations for the Events with Moment Magnitude Smaller than* -2, Bull. Seismol. Soc. Am. *81*, 1157–1182.

GRIFFITH, A. A. (1921), *The Phenomena of Rupture and Flow in Solids*, Phil. Trans. Roy. Soc. of London *A221*, 163–197.

GRIFFITH, A. A. (1924), *The Theory of Rupture*, Proc. Ist Int. Conf. Applied Mech., Delft, 55–93.

HOEK, E. (1965), *Rock Fracture under Static Stress Conditions*, CSIR Report MEG 383, National Mechanical Eng. Research Institute, Council for Scientific and Industrial Research, Pretoria, South Africa.

HOEK, E. *Brittle failure of rocks in rock mechanics in engineering practice*. In *Rock Mechanics in Engineering Practice* (eds. Stagg, K. G., and Zienkiewicz, O. C.) (John Willey and Sons 1968) pp. 99–124.

HORII, H., and NEMAT-NASSER, S. (1985), *Compression-induced Microcrack Growth in Brittle Solids: Axial Splitting and Shear Failure*, J. Geophys. Res. *90* (B4), 3105–3125.

KAISER, P. K., and MALONEY, S. M. (1997), *Scaling Laws for the Design of Rock Support*, Pure appl. geophys. *150*, 415–434.

LAJTAI, E. Z., CARTER, B. J., and AYARI, M. L. (1990), *Criteria for Brittle Fracture in Compression*, Engin. Fract. Mech. *37* (1), 25–49.

LEE, M. Y., and HAIMSON, B. C. (1993), *Laboratory Study of Borehole Breakouts in Lac du Bonnet Granite: A Case of Extensile Failure Mechanism*, Int. J. Rock Mech. Min. Sci. and Geomech. Abstr. *30* (7), 1039–1045.

LOCKNER, D. A., BYERLEE, J. D., KUKSENKO, V., PONOMAREV, A., and SIDORIN, A., *Observation of quasi-static fault growth from acoustic emissions*. In *Fault Mechanics and Transport Properties of Rocks* (eds. EVANS, B., and WONG, T.-f.) (Academic Press 1992) pp. 3–31.

MADARIAGA, R. (1976), *Dynamics of an Expending Circular Fault*, Bull. Seismol. Soc. Am. *66*, 639–666.

MARTIN, C. D. (1997), *Seventeenth Canadian Geotechnical Colloquium: The Effect of Cohesion Loss and Stress Path on Brittle Rock Strength*, Canadian Geotech. J. *34* (5), 698–725.

MARTIN, C. D., and CHANDLER, N. A. (1994), *The Progressive Fracture of Lac du Bonnet Granite*, Int. J. Rock Mech. Min. Sci. *31*, 643–659.

MARTIN, C. D., KAISER P. K., and McCREATH, D. R. (1998), *Hoek-Brown Parameters for Predicting the Depth of Brittle Failure around Tunnels*, Canadian Geotechnical Journal, accepted for publication in 1999 Feb. issue.

MARTIN, C. D., and READ, R. S. (1996), *AECL's Mine-by Experiment: A Test Tunnel in Brittle Rock*, Proc. 2nd North American Rock Mech. Symposium (eds. Aubertin, Hassani and Mitri), *1*, 13–24.

MARTIN, C. D., READ, R. S., and MARTINO, J. B. (1997), *Observation of Brittle Failure around a Circular Test Tunnel*, Int. J. Rock Mech. Min. Sci. *34*, 1065–1073.

McGARR, A., *Some applications of seismic source mechanism studies to assessing underground hazard*. In *Rockburst and Seismicity in Mines* (eds. Gay, N. C., and Wainwright, E. H.), Symp. Ser. *6* (South Africa Inst. Min. Metal 1984) pp. 199–208.

MENDECKI, A. J., *Quantitative seismology and rock mass stability*. In *Seismic Monitoring in Mines* (ed. Mendecki, A. J.) (Chapman and Hall, London 1997) Chapter 10, pp. 178–219.

MYER, L. R. KEMENY, J. M., ZHENG, Z., SUAREZ, R., EWY, R. T., and COOK, N. G. W. (1992), *Extensive Cracking in Porous Rock under Differential Compressive Stress*, Appl. Mech. Rev. *45*, 263–280.

OLSSON, O. L., and WINBERG, A., *Current understanding of extent and properties of the excavation disturbed zone and its dependence of excavation method*, In *Proc. Int. Conf. on Deep Geological Disposal of Radioactive Waste* (eds. Martino, J. B., and Martin, C. D.) (Canadian Nuclear Society, Toronto 1996) pp. 101–112.

ORTLEPP, W. D., *Rock Fracture and Rockbursts—An Illustrative Study*, Monograph Series M9 (The South African Institute of Mining and Metallurgy, Johannesburg, 1997).

RANDALL, M. J. (1973), *The Spectral Theory of Seismic Source*, Bull. Seismol. Soc. Am. *63*, 1133–1144.

READ, R. S., and MARTIN, C. D. (1996), *Technical Summary of AECL's Mine-by Experiment, Phase 1: Excavation Response*, AECL (AECL-11311, COG-95-171).

REVALOR, R., JOSIEN, J. P., BESSON, J. L., and MAGRON, A., *Seismic and seismoacoustic experiments applied to the prediction of rockbursts in French coal mines*. In *Rockbursts and Seismicity in Mines* (ed. C. Fairhurst) (Balkema, Rotterdam 1990) pp. 301–306.

SATO, T. (1978), *A Note on Body-wave Radiation from Expanding Tension Crack*, Sci. Rep. Tohoku Univ., Ser. 5, Geophys. *25*, 1–10.

SCHOLZ, C. H., *The Mechanics of Earthquakes and Faulting* (Cambridge University Press 1990).

STACEY, T. R. (1981), *A Simple Extension Strain Criterion for Fracture of Brittle Rock*, Int. J. Rock Mech. Min. Sci. and Geomech. Abstr. *18*, 469–474.

TALEBI, S. (1993), *Source Studies of Mine-induced Seismic Events over a Broad Magnitude Range* $(-4 < M < 4)$, CANMET, Report MRL 93–046 (CL).

TALEBI, S., and YOUNG, R. P., *Failure Mechanism of Crack Propagation Induced by Shaft Excavation at the Underground Research Laboratory*. In *Proc. Int. Conf. on Rock Mech. and Rock Physics at Great Depth* (eds. Maury, V., and Fourmaintraux, D.) (A. A. Balkema, Rotterdam 1989), *2*, 719–726.

TALEBI, S., and YOUNG, R. P. (1992), *Microseismic Monitoring in Highly Stressed Granite: Relation between Shaft-wall Cracking and in situ Stress*, Int. J. Rock Mech. Min. Sci. and Geomech. Abstr. *29*, 25–34.

TRUIFU, C. I., URBANCIC, T. I., and YOUNG, R. P. (1995), *Source Parameters of Mining-induced Seismic Event: An Evaluation of Homogeneous and Inhomogeneous Faulting Model for Assessing Damage Potential*, Pure appl. geophys. *145* (1), 3–27.

URBANCIC, T. J., YOUNG, R. P., BIRD, S., and BAWDEN, W. (1992), *Microseismic source parameters and their use in characterizing rock mass behavior: Consideration from Strathcona mine*, In *Proc. AGM-CIM*, May, 36–47.

WILES, T., *Map 3-D User's Manual* (Mine Modeling Limited 1996).

WONG, T. F. (1982), *Micromechanics of Faulting in Westerly Granite*, Int. J. Rock Mech. Min. Sci. and Geomech. Abstr. *19*, 49–64.

YOUNG, R. P. (1993), *Seismic Methods Applied to Rock Mechanics*, ISRM News J. *1* (3), 4–18.

(Received March 9, 1998, accepted July 24, 1998)

Seismicity Induced by Fluid Injections

Pure appl. geophys. 153 (1998) 95–111
0033–4553/98/010095–17 $ 1.50 + 0.20/0

Pure and Applied Geophysics

Injection-induced Microseismicity in Colorado Shales

SHAHRIAR TALEBI,[1] TOM J. BOONE[2] and JOHN E. EASTWOOD[3]

Abstract—Imperial Oil Resources Limited uses cyclic steam stimulation to recover oil from their Cold Lake oil field in Alberta. This operation, in particular situations, can be associated with the failure of well casings in the Colorado shales above the oil-bearing formation. A number of fluid injection operations was undertaken at this site and the associated microseismicity was detected using two three-component geophones and fifteen hydrophones. The purpose of this experiment was to simulate the occurrence of a casing failure, determine the feasibility of monitoring in a shallow environment, and characterize the microseismic activity. A calibration survey provided values of 1786 ± 108 m/s for P-wave velocity, 643 ± 56 m/s for S-wave velocity and 0.428 ± 0.017 for Poisson's ratio in the shale formation. Estimates of the quality factor Q_P were 15 for the horizontal direction and 38 for the vertical direction, corroborating the evidence of velocity anisotropy. Calibration shots were located to within 10 m of the actual shot points using triangulation and polarization techniques. Several hundred microseismic events were recorded and 135 events were located. The results showed that microseismic activity was confined to depths within 10 meters of the injection depth. The experiment clearly established the feasibility of detecting microseismicity induced by fluid injection rates typical of casing failures in shales at distances over 100 m.

Key words: Water injection, microseismicity, source location, P- and S-wave velocity, attenuation, polarization.

Introduction

Injection of fluids under high pressure into a rock mass has several scientific and industrial applications, including the measurement of regional stress fields and the development of deep exchangers in hot-dry-rock geothermal projects. In the oil industry, fluid injections are mostly undertaken in order to increase the productivity of oil wells. Among different techniques investigated to detect fracture propagation during fluid injection operations in rock masses, monitoring and analysis of microseismic activity associated with fluid percolation has been proved to be the most

[1] CANMET, 1079 Kelly Lake Rd., Sudbury, Ontario, Canada P3E 5P5. Fax: (705) 670-6556, E-mail: stalebi@nrcan.gc.ca
[2] Imperial Oil Resources Ltd., 3535 Research Rd. NW, Calgary, Alberta, Canada T2L 2K8.
[3] Exxon Production Research Company, P.O. Box 2189, Houston, Texas, 77252, U.S.A.

reliable technique. This method has been in use over the last two decades in geothermal applications (e.g., ALBRIGHT and PEARSON, 1980; PEARSON, 1981; CASH *et al.*, 1983; NIITSUMA *et al.*, 1985; MURPHY and FEHLER, 1986; TALEBI and CORNET, 1987; HOUSE, 1987) as well as oil and gas applications (e.g., POWER *et al.*, 1976; THORNE and MORRIS, 1988; SARDA *et al.*, 1988; TALEBI *et al.*, 1991; DEFLANDRE and DUBESSET, 1992). Because of technical limitations and the cost of the application of this technique at large depths, many authors have been limited to the use of only one borehole for sensor installation and sometimes even only one three-component sensor. Determination of the geometry of hydraulic fractures induced by fluid injections has been the dominant objective of the application of this technique in the oil industry.

This paper summarizes the results of a pilot project, based on the use of microseismic techniques in the Cold Lake oil field in Alberta. The extremely high viscosity of the bitumen in this oil sands deposit is the main obstacle for economic recovery of heavy oil. The procedure used for oil extraction is Cyclic Steam Stimulation (CSS) which consists of injecting large volumes of steam under high pressure and temperature (300°C) to reduce the viscosity of the bitumen and allow its flow towards producing wells. Steam injection is alternated with oil and water production for as many cycles as economic conditions permit (KRY, 1989). The oil-bearing Clearwater formation, located 400–450 m below surface, is the target of these operations. In particular situations, this operation can cause failure of well casings in the 200–250 m depth-range in the Colorado shale formation, which causes concern from an operational and environmental point of view. The development of a reliable tool capable of early detection of casing failures is of utmost importance to the operation of this oil field.

The pilot project was designed to simulate the occurrence of a casing failure in the shale formation and to detect the associated microseismic activity. Water injection operations at different flow rates were undertaken at a depth of 220 m in a location remote from oil-producing wells. A calibration survey was performed in order to measure the seismic properties of the shale formation. The main objective of this project was to determine if microseismic monitoring techniques could constitute an effective early-warning tool for detecting fluid flow caused by casing failures. Other issues to be addressed by the project consisted of the comparison of different source location techniques, determination of maximum distances at which microseismic events can be reliably detected and the design of an optimum sensor array for a large-scale application of the technique.

Description of the Experiment

Figure 1 shows a plan view of the test site and the location of the injection well and the four observation wells. Two three-component geophones with dual-gain preamplifiers programmable from the surface were specially designed for this experiment and cemented in the borehole (G) at depths of 199 m ($G1$) and 249 m ($G2$). The three remaining observation wells ($H1$, $H2$ and $H3$) were used for the installation of strings of five hydrophones (depths of 180 m, 200 m, 220 m, 240 m and 260 m). The key issue in the design of the geometry of the array of sensors was to provide suitable coverage of the area around the injection point, which was expected to generate microseismic activity. Signals from the two three-component geophone assemblies were amplified down hole at two different gains and passed through the conditioning and triggering boards of a data acquisition system, along with signals from the three hydrophone strings. The A/D boards acquired data whenever user-defined criteria on triggering conditions were met. Calibration tests revealed a flat frequency response of the entire system up to 1.5 kHz. The sampling rate was 7.5 kHz/channel and the anti-aliasing filters with a slope of 72 dB/octave were set at 2 kHz.

Water injection operations took place in four stages during September and October 1995 (Table 1). The level of the aquifer was monitored throughout the experiment using a borehole at a distance of about 120 m from the injection well. Since the nearest aquifer was located at depths of 100–120 m above the injection point, it was expected that the injected water would have little effect on the aquifer level. The observations, however, showed that there was very rapid communication with the aquifer, as indicated by the strong correlation between the aquifer level below surface and the total injected volume (see Fig. 2). Possible causes of this occurrence include fluid flow along a pre-existing vertical fracture or along the

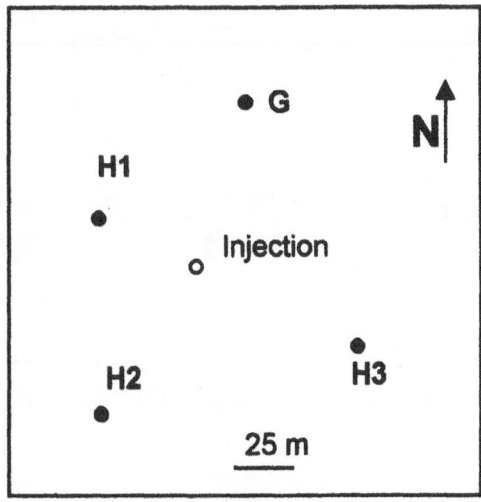

Figure 1
Plan view of the test site showing the location of the injection well and the boreholes used to install three-component geophones (G) and string of five hydrophones ($H1$, $H2$ and $H3$).

Table 1

Summary of the four stages of injection operations

Stage	Start date	Duration (hours:min.)	Average rate (m³/hr)	Injected volume (m³)
1	Sept. 14, 1995	71:50	2.7	191
2	Sept. 18, 1995	20:39	10.2	208
3	Oct. 2, 1995	11:20	28.2	320
4	Oct. 10, 1995	17:08	24.3	416

interfaces between the casing, cement and the rock mass. For the purposes of the pilot project, this observation simply means that only a fraction of the injected fluids was entering the shales so that the flow rates and injected volumes reported in Table 1 are upper bound values. Aquifer monitoring can be an alternative method of casing failure detection provided there is immediate communication to an aquifer that effectively behaves as a confined volume. In general, however, immediate communication to the aquifer is not the case for a typical casing failure and reliable interpretation of such data can be difficult due to a number of factors including the size and permeability of the aquifer and the natural rise and fall of its level.

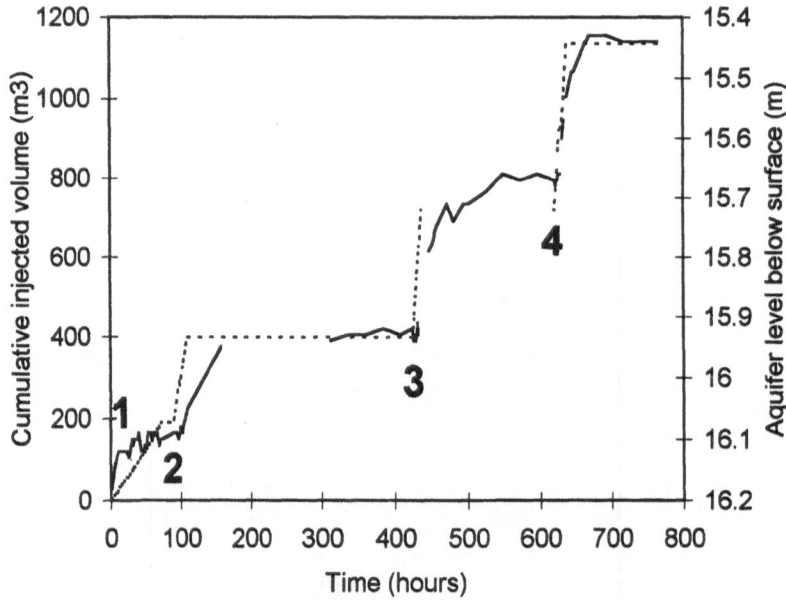

Figure 2

Comparison between variations, as a function of time, of the aquifer level in meters below surface (solid line), and cumulative injected volume in m³ (broken line). Bold numbers indicate start times of the four stages of injection operation.

During stage 1 of injection operation the down-hole gain of geophones was gradually increased to 84 dB and 104 dB on low-gain and high-gain sensors respectively, and the gains of hydrophones on surface reached 34 dB to 40 dB. The triggering parameters were set to detect any events barely emanating from the background noise. However, no significant microseismic activity was detected during stages 1 and 2 of the operation. Much higher flow rates were used in stages 3 and 4 (Table 1). Low-frequency noise levels on geophone sensors increased and microseismic events first were recorded early in stage 3. This trend continued during stage 4 as events of substantially higher quality were detected early in the operation and continued to be recorded throughout this stage.

Velocity and Attenuation Measurements

Sonic logs from three wells in the vicinity showed average values of 2017 m/s and 667 m/s for P- and S-wave velocities between 180 m and 250 m of depth. The average Poisson's ratio was 0.425, 0.439 and 0.446 for the three boreholes. Apart from providing a range for elastic properties for the shale, sonic logs exhibited variations of these properties as a function of depth and well location. In order to measure these parameters in the frequency range of monitoring, six primacord shots were detonated in the injection well and one shot was fired in $H3$ between 180 m and 220 m of depth (Fig. 1). The exact time of shot detonation was recorded on-site for the first four shots performed prior to injection operations. P-wave arrivals were very clear on all sensors and generally easy to pick due to high signal-to-noise ratios. The polarity of the first arrival of P waves was clearly compressional on all recorded signals for all the shots (Fig. 3). S-wave arrivals, however, were difficult to pick particularly for hydrophone signals. The average results for these four shots can be summarized as follows:

$$P\text{-wave velocity:} \quad 1786 \pm 108 \text{ m/s}$$

$$S\text{-wave velocity:} \quad 643 \pm 56 \text{ m/s}$$

$$\text{Poisson's ratio:} \quad 0.428 \pm 0.017.$$

Shales are generally considered to be transversely anisotropic (e.g., JOHNSTON and TOKSÖZ, 1980; THOMSEN, 1986; WHITE et al., 1982). The symmetry axis is usually the vertical axis and the elastic medium is characterized by five independent constants. In this case, which is equivalent to the case of stratified media with horizontal layering, S waves decouple into SH and SV components. The velocity model here is sometimes considered to be a spheroid; i.e., an ellipsoid with one axis parallel to the vertical direction and the two other axes of a similar size in the horizontal plane. Figure 4 shows the projection of P-wave velocity results on horizontal and vertical axes. A similar approach was not attempted for S waves

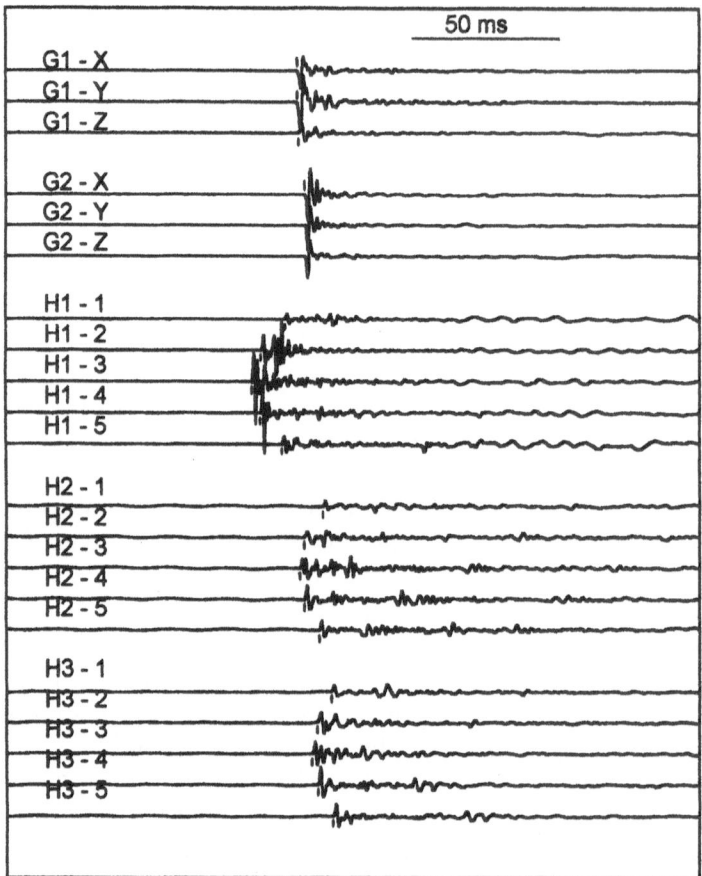

Figure 3
Signals from the first calibration shot at 220 m of depth.

because of the insufficient number of measurements. The best-fit ellipse to the
P-wave data indicates a velocity anisotropy of 10.4%, defined as $(V_{PH} - V_{PZ})/V_{PH}$
where V_{PH} and V_{PZ} are velocities in horizontal and vertical directions. The large
variation around the horizontal direction seems to indicate the predominance of the
effects of layering on the results.

Attenuation measurements were attempted, using the rise time method originally proposed by GLADWIN and STACEY (1974) based on an empirical
relationship:

$$\tau = \tau_0 + CT/Q, \tag{1}$$

where τ is the rise time of the recorded signal, τ_0 is the rise time of the source, C
is a constant, T is travel time and Q is the quality factor of the medium. According
to KJARTANSSON (1979), who provided a theoretical basis for the above relation-

ship for impulsive sources, C can only be considered to be a constant when Q is larger than 20. The asymptotic values of this parameter are 0.485 and 0.298 for displacement and velocity signals, respectively. BLAIR and SPATHIS (1982) have indicated that the above equation is strictly valid for impulsive sources ($\tau_0 = 0$) however problems arise for realistic sources, as C becomes dependent on the properties of the source. BLAIR (1982) points out the effect of the frequency response of the system on the results and the difficulty of reliable measurements, especially within 10 m of an impulsive source. STEWART (1984) considered different types of non-impulsive sources and showed that although the proportionality of rise time and attenuation is valid only for an impulsive source function, the maximum attenuation experienced by a seismic pulse in the source-sensor path can still be estimated using this method. Some authors have reported comparable results using the rise time method and the spectral ratio method (JANNSEN *et al.*, 1985; BOURBIÉ *et al.*, 1986).

Figure 5a shows a plot of the rise times versus travel times for P-wave signals detected on the two geophones from the six shots in the injection well. The straight line corresponds to a quality factor of 32 obtained by MCDONAL *et al.* (1958) from *in situ* measurements in Pierre shale, a formation that has comparable seismic properties to the shale formation considered here. The straight line seems to provide a lower bound to our data although the scatter indicates a possible dependence on the angle of incidence. We assume a value of 0.1 ms for τ_0 based on previous experiments using the same type of source (TALEBI *et al.*, 1991), and

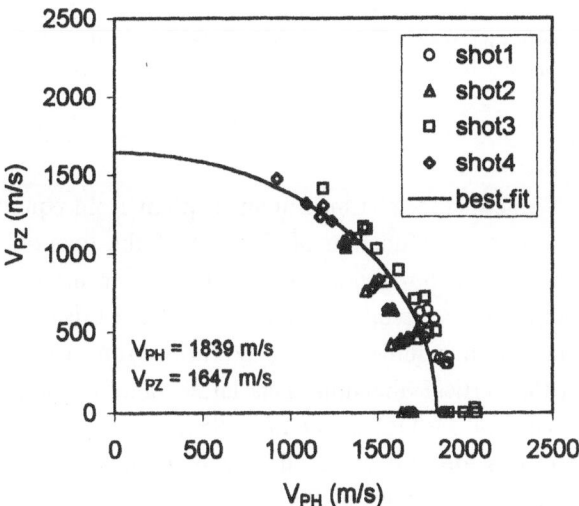

Figure 4
P-wave velocities projected on horizontal and vertical axes.

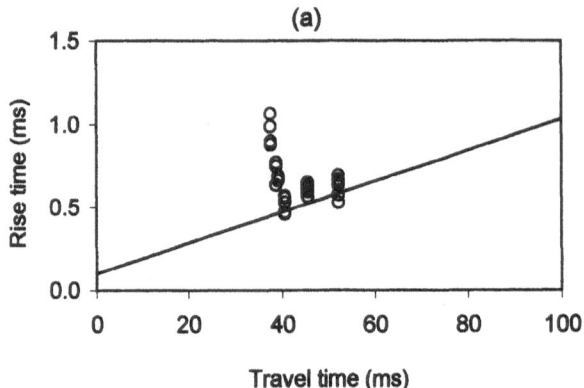

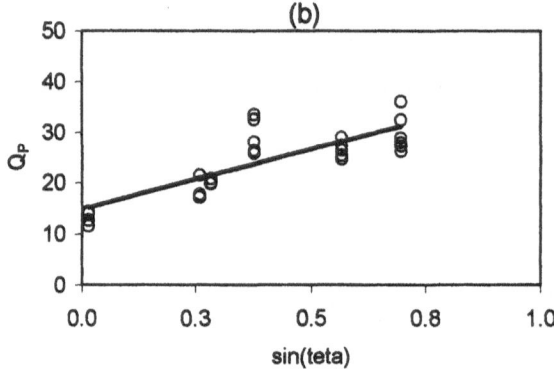

Figure 5
(a) Rise time versus travel time for *P* waves. The straight line shows the relationship expected for $Q_P = 32$; (b) Q_P versus the sine of the take-off angle between the source-sensor rays and the horizontal plane.

calculate the quality factor for each measurement point from equation (1). Figure 5b illustrates the results as a function of the sine of the take-off angle between source-sensor rays and the horizontal plane. The scatter in the data has been considerably reduced and the results strongly indicate the existence of attenuation anisotropy. The best-fit line indicates that Q varies from 15 in the horizontal direction to 38 in the vertical direction. This latter value is comparable with the value of 32 obtained by McDONAL *et al.* (1958) for the vertical direction in Pierre shale. The results are also compatible with those of JOHNSTON and TOKsöz (1980) for Colorado oil shales in the laboratory where Q_P was estimated to be about 28 and 14 perpendicular and parallel to the bedding planes, respectively.

Sensor Orientation

The seven calibration shots were used to estimate the orientation of different components of the two geophones. In the first step, the polarization angles of the P-wave signals were calculated; i.e., two rotation angles needed to put all the energy along the X component of each geophone (e.g., MATSUMURA, 1981). No assumptions were made about the plunge of the Z components which were estimated independently as a check to the accuracy of the technique. Sensor orientations were then calculated given the coordinates of shot and geophone locations. Table 2 gives the estimated azimuth of the X components and plunge of the Z components of the two geophones. The plunges of the Z component of geophone 1 revealed discrepancies and were corrected assuming a layer with a slightly different velocity at the location of this sensor. The convention for rotations is a left-handed system and angles are positive for clockwise rotations around the positive Z (vertical) and Y axes of the sensors. If we consider the average values for shots 2–7, the orientation of Z components can be estimated to within 1° of the vertical direction. The accuracy should be better for azimuth determinations because a smaller scatter is observed. Individual measurements, however, can be off by $\pm 1°$ for azimuth and $\pm 5°$ for plunge determinations. The overall results indicate that accurate sensor orientation can be made with cemented sensors, using an adequate calibration survey. The lower accuracy in the determination of the plunge of Z components is compatible with the layered nature of the shales and previous results on seismic anisotropy.

Locations of calibration shots were estimated by applying triangulation and polarization methods. The triangulation method was based on a damped least-squares technique using P- and S-wave arrival times. The results are shown in Figure 6 using two different symbols for (tp) indicating that only P-wave arrival times were used, and for $(tp\&ts)$ indicating that both P- and S-wave arrival times were used. Another method consisted of analyzing the polarization of the incident P waves in order to determine the source-sensor direction and the time difference between P- and S-wave first arrivals used for calculating source-sensor distances. Shot locations could then be estimated independently from signals of the top

Table 2

Azimuth of the X components and plunge of the Z components of the two geophones

Geophone	Shot(s)	Borehole	Depth (m)	X (deg.)	Z (deg.)
1	1	$H3$	220	137.1	3.8
1	2–7	Injection	180–220	132.5 ± 0.6	-0.5 ± 2.9
2	1	$H3$	220	171.1	-4.8
2	2–7	Injection	180–220	167.1 ± 1.0	-0.9 ± 3.8

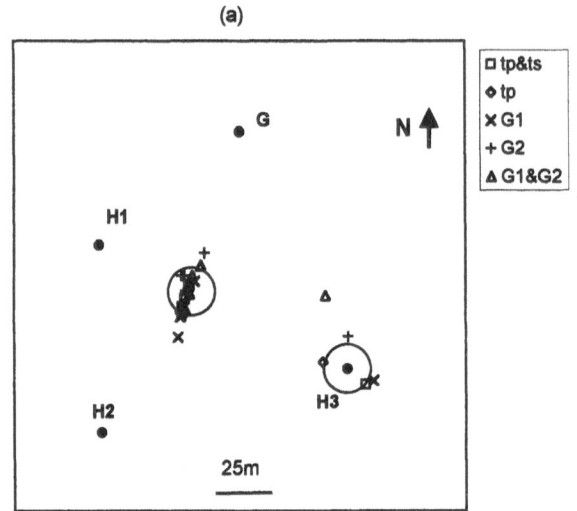

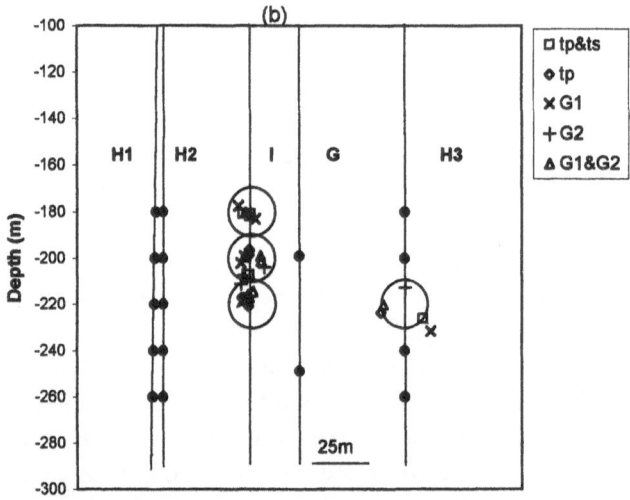

Figure 6
Plan view (a) and cross section along E–W (b) of source location results of calibration shots using
different methods. Solid circles show sensor locations and open circles show a sphere of 10 m radius
centered on shot locations. The horizontal and vertical scales are identical.

geophone (*G*1) and the bottom geophone (*G*2) (see Fig. 6). The last method was
based on using only the polarization direction of incident *P* waves on the two
geophones. The results of this method, indicated by (*G*1&*G*2) in Figure 6, were
obtained by calculating the point of intersection of the incident rays on the two
geophones. The lower accuracy of plunge compared to azimuth estimations,

discussed in the previous paragraph, manifests itself in the horizontal plane as hypocenters from polarization methods tend to be aligned along source-sensor directions. The overall results showed that reasonable estimates within ± 10 m of the actual shot locations could be made in the central area utilizing most of these methods.

Microseismicity

Microseismic events were first detected shortly after the beginning of high flow-rate injection tests (stage 3). It took a few hours before these events were strong enough to be locatable and about ten hours before a reasonable data set was collected. At the end of the injection operations (stage 4) several hundred microseis-

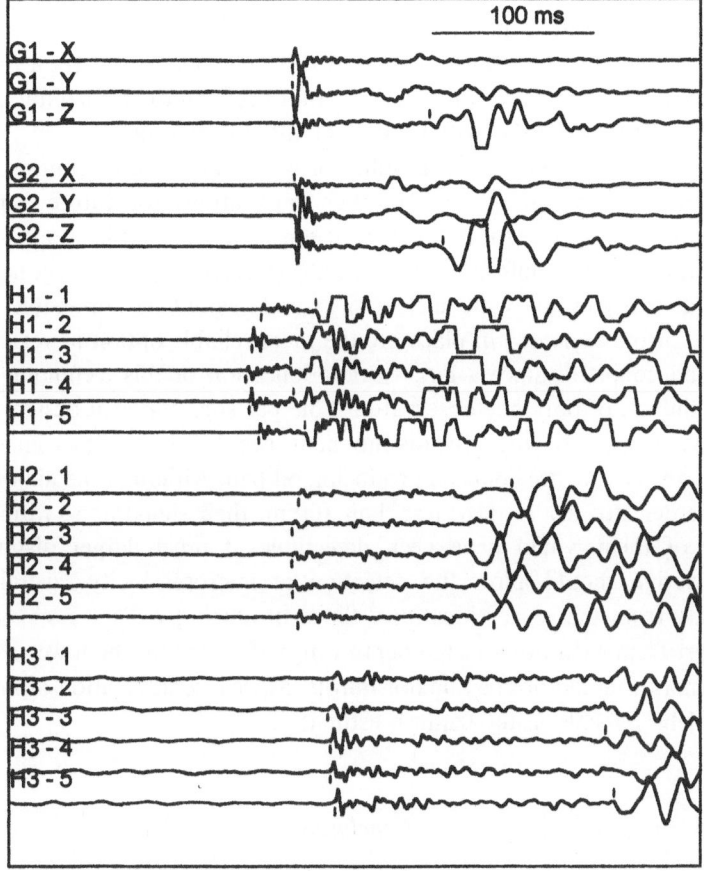

Figure 7
Signals of the largest microseismic event recorded on October 2, 1995.

mic events were recorded. Figure 7 illustrates signals from the largest recorded event, showing very clear onsets of P and S waves. A few other events had clear P and S arrivals on geophone and hydrophone sensors and could be located using first arrival times. The accuracy of source location determinations for these events was estimated to be about 10 m by comparing the results of different methods. For the majority of the events, however, the signal-to-noise ratio on hydrophone signals was too low to allow a clear estimation of P- and S-wave arrival times. A total of 135 events had a sufficient signal-to-noise ratio on the geophone sensors and could be located using polarization methods. Because of the difficulty of the accurate picking of S-waves onsets, particularly towards the end of the experiment, the most reliable results were obtained using the orientation of the incident P-wave signals alone. Figures 8, 9 and 10 show source location results for three time periods during the two high-flow rate injection operations (stages 3 and 4). With the exception of one event, all the activity clusters within a few tens of meters around the injection point. In the horizontal plane, the apparent progression along the line between the injection hole and well G results from the uncertainty in plunge determinations discussed in the previous section. In spite of this bias, one can detect a progression in the extent of the seismic region as a function of time in the horizontal plane. The activity is confined to depths within 10 m of the injection point.

The present results show the feasibility of casing failure detection using microseismic techniques, but have several other implications for future applications. Firstly, the use of only one three-component sensor in such applications is not ideal as one would often be confronted with the uncertainty in phase identification and the fact that many events with unclear S arrivals would be impossible to locate. Geophone sensors cemented in place are the most reliable option as regards sensor selection and coupling. Since a large-scale application of this technique would be limited to the use of only one observation hole per site, our final design for future applications consists of five three-component geophone sensors cemented in a borehole in the central area of each producing oil pad. Although the majority of the events were detected at distances less than 100 m, their signal/noise ratios indicate that most of them would have been detectable at much larger distances. The proposed design should allow the detection of microseismicity several hundred meters away from the sensor locations. Since producing wells are several tens of meters apart from each other at the depth range 200–250 m, the lower accuracy in plunge estimates should not be a major hurdle, as clear identification of any leaking well should be possible using azimuth estimates.

Conclusion

The calibration survey provided values of 1786 ± 108 m/s for P-wave velocity, 643 ± 56 m/s for S-wave velocity and 0.428 ± 0.017 for Poisson's ratio in the shale

(a)

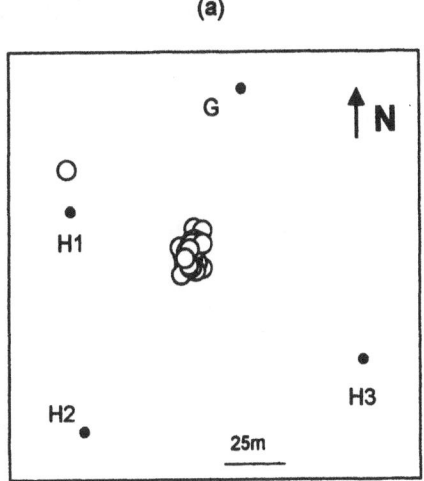

(b)

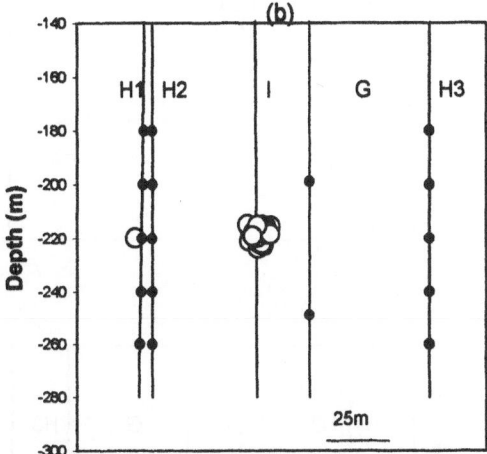

Figure 8
Plan view (a) and cross section along E–W (b) of location of events recorded on October 2, 1995. The horizontal and vertical scales are identical.

formation. Estimates of the quality factor Q_P were 15 for horizontal direction and 38 for vertical directions, corroborating the evidence of velocity anisotropy. The orientations of Z components of the two three-component sensors were estimated to within 1° of the vertical direction. Individual measurements, however, could be off by ±1° for azimuth and ±5° for plunge determinations. The overall results showed that accurate sensor orientation can be made with cemented sensors using an adequate calibration survey. Also, estimates of actual shot locations to within 10 m can be made in the central area employing different methods.

The location of microseismic events showed that the activity originated from an area close to the injection point, but remained confined to depths within 10 m of the injection depth. Although the majority of the events were detected at distances less than 100 m, their signal/noise ratios indicated that most of them would have been detectable at considerably longer distances. The experiment clearly established the feasibility of a large-scale application of microseismic technology as a detection

(a)

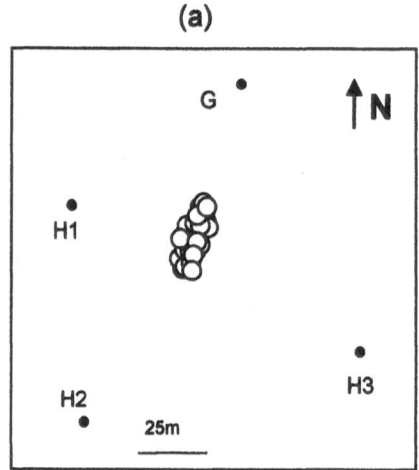

(b)

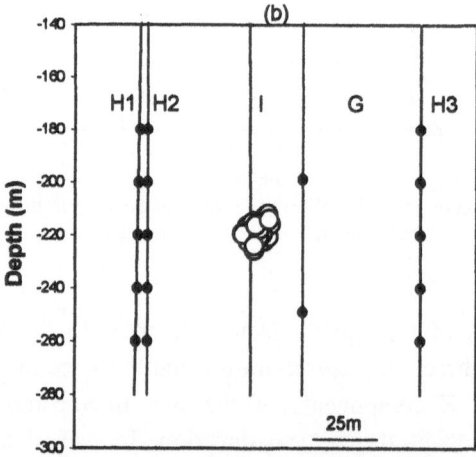

Figure 9
Plan view (a) and cross section along E–W (b) of location of events recorded on October 10, 1995. The horizontal and vertical scales are identical.

(a)

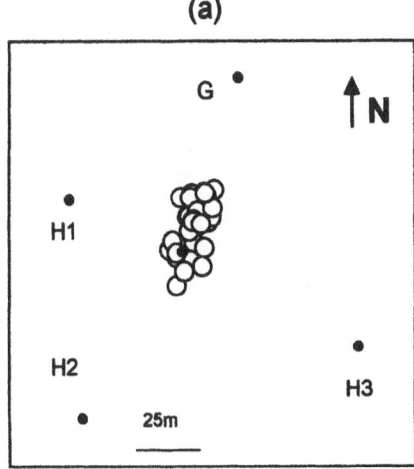

(b)

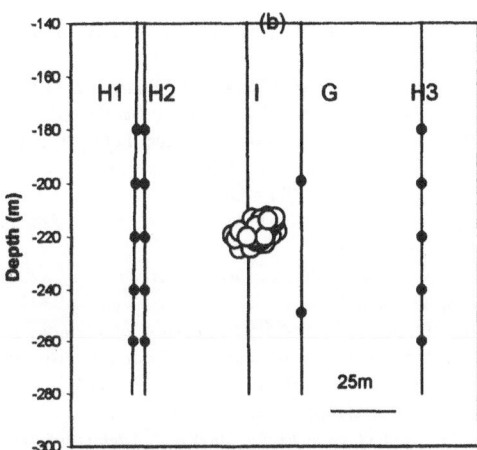

Figure 10
Plan view (a) and cross section along E–W (b) of location of events recorded on October 11, 1995. The horizontal and vertical scales are identical.

tool for casing failures in oil sands. Since a large-scale application of this technique would be limited to the use of only one observation hole per site, the design for future applications consists of five three-component geophone sensors cemented in a borehole in the central area of each producing oil pad. Such a design should allow the detection of microseismicity several hundred meters distant.

Acknowledgments

The authors would like to thank Imperial Oil Limited, Resources Division, for their financial support of this project and their permission to publish the data. The collaboration of our colleagues in IOL (Rick Bailey, Karl Pustanyk, Darcy Ward) and CANMET (Doug Becker, Denis Lebel, Paul Rochon, Parviz Mottahed) was instrumental in the successful completion of this project. We are grateful to S. J. Gibowicz and A. McGarr for their careful reviews which enhanced the quality of this paper.

REFERENCES

ALBRIGHT, J. N., and PEARSON, C. F. (1980), *Location of Hydraulic Fractures Using Microseismic Techniques*, Proc. Ann. Tech. Conf., Soc. Pet. Eng., Dallas, Texas, Sept. 21–24, SPE Paper 9509.

BLAIR, D. P. (1982), *Measurement of Rise Times of Seismic Pulses in Rock*, Geophysics 47, 1047–1058.

BLAIR, D. P., and SPATHIS, A. T. (1982), *Attenuation of Explosion-generated Pulse in Rock Masses*, J. Geophys. Res. 87, 3885–3892.

BOURBIÉ, T., COUSSY, O., and ZINSZNER, B., *Acoustique des Milieux Poreux* (éditions Technip, Paris 1986).

CASH, D., HOMUTH, E. F., KEPPLER, H., PEARSON, C., and SASAKI, S. (1983), *Fault-plane Solutions for Microearthquakes Induced at the Fenton Hill Hot Dry Rock Geothermal Site: Implication for the State of Stress near a Quaternary Volcanic Center*, Geophys. Res. Lett. 10, 1141–1144.

DEFLANDRE, J. P., and DUBESSET, M. (1992), *Identification of P/S Successions for Application in Microseismicity*, Pure appl. geophys. 139, 405–420.

HOUSE, L. (1987), *Locating Microearthquakes Induced by Hydraulic Fracturing in Crystalline Rock*, Geophys. Res. Lett. 14, 919–921.

GLADWIN, M. T., and STACEY, F. D. (1974), *Anelastic Degradation of Acoustic Pulses in Rock*, Phys. Earth and Planet. Inter. 8, 332–336.

JANNSEN, D., VOSS, J., and THEILEN, F. (1985), *Comparison of Methods to Determine Q in Shallow Marine Sediments from Vertical Reflection Seismograms*, Geophys. Prospect. 33, 479–497.

JOHNSTON, D. H., and TOKSÖZ, M. N. (1980), *Ultrasonic P- and S-wave Attenuation in Dry and Saturated Rocks under Pressure*, J. Geophys. Res. 85, 925–936.

KJARTANSSON, E. (1979), *Constant Q-wave Propagation and Attenuation*, J. Geophys. Res. 84, 4737–4748.

KRY, P. R. (1989), *Field Observations of Steam Distribution during Injection to the Cold Lake Reservoir*, Proc. Symp. Rock at Great Depth, Pau, France, 853–861.

MATSUMURA, S. (1981), *Three-dimensional Expression of Seismic Particle Motions by the Trajectory Ellipsoid and its Application to the Seismic Data Observed in the Kanto District, Japan*, J. Phys. Earth 29, 221–239.

MCDONAL, F. J., ANGONA, F. A., MILLS, R. L., SENGBUSH, R. L., VAN NOSTRAND, R. G., and WHITE, J. E. (1958), *Attenuation of Shear and Compressional Waves in Pierre Shale*, Geophysics 23, 421–439.

MURPHY, H. D., and FEHLER, M. C. (1986), *Hydraulic Fracturing of Jointed Formations*, Proc. Int. Meeting on Pet. Eng., Soc. Pet. Eng., Beijing, March 17–20, SPE Paper 14088.

NIITSUMA, H., NAKATSUKA, K., CHUBACHI, N., YOKOYAMA, H., and TAKANOHASHI, M. (1985), *Downhole AE Measurement of a Geothermal Reservoir and its Application to Reservoir Control*, Proc. 4th Conf. AE/MS in Geol. Struc. and Mat., Trans-Tech Publications, 475–489.

PEARSON, C. (1981), *The Relationship between Microseismicity and High Pore Pressures during Hydraulic Stimulation Experiments in Low Permeability Granitic Rocks*, J. Geophys. Res. 86, 7855–7864.

POWER, D. V., SCHUSTER, C. L., HAY, R., and TWOMBLY, J. (1976), *Detection of Hydraulic Fracture Orientation and Dimensions in Cased Wells*, J. Pet. Tech., 1116–1124.

SARDA, J. P., PERREAU, P. J., and DEFLANDRE, J. P. (1988), *Acoustic Emission Interpretation for Estimating Hydraulic Fracture Extent: Laboratory and Field Studies*, Proc. 63rd Ann. Tech. Conf., Soc. Pet. Eng., Houston, Texas, October 2–5, SPE Paper 18192.

STEWART, R. C. (1984), *Q and the Rise and Fall of a Seismic Pulse*, Geophys. J. R. Astr. Soc. *76*, 793–805.

TALEBI, S., and CORNET, F. H. (1987), *Analysis of the Microseismicity Induced by a Fluid Injection in a Granitic Rock Mass*, Geophys. Res. Lett. *14*, 227–230.

TALEBI, S., VANDAMME, L., MCGAUGHEY, W. J., and YOUNG, R. P., *Microseismic mapping of a hydraulic fracture. In Rock Mechanics as a Multidisciplinary Science* (ed. Roegiers, J. C.) (Balkema, Rotterdam 1991) pp. 461–470.

THOMSEN, L. (1986), *Weak Anisotropy*, Geophysics *51*, 1954–1966.

THORNE, B. J., and MORRIS, H. E. (1988), *An Assessment of Borehole Seismic Fracture Diagnostics*, Proc. 63rd Ann. Tech. Conf., Soc. Pet. Eng., Houston, Texas, October 2–5, SPE Paper 18193.

WHITE, J. E., MARTINEAU-NICOLETIS, I., and MONASH, C. (1982), *Measured Anisotropy in Pierre Shale*, Geophys. Prosp. *31*, 709–725.

(Received June 12, 1998, revised/accepted August 10, 1998)

Pure appl. geophys. 153 (1998) 113–130
0033–4553/98/010113–18 $ 1.50 + 0.20/0

Pure and Applied Geophysics

Source Parameters of Injection-induced Microseismicity

SHAHRIAR TALEBI[1] and TOM J. BOONE[2]

Abstract—We analyze source parameters of microseismic events ($M < -1$) associated with high flow-rate water injections in a shale formation at a depth of 220 m. Two types of events were observed: several hundred impulsive events with clear *P*- and *S*-wave arrivals, and continuous emissions with peaked spectra detected well into the experiment. For a representative collection of impulsive events, an ω^{-2} model provided satisfactory fits to displacement spectra corrected for attenuation, and average quality factors of 34 and 15 were obtained for *P* and *S* waves. *P*-wave first motion analysis and E_S/E_P ratios indicated the existence of a non-double-couple component in some events, particularly early in the experiment. A clear difference was observed for estimates of stress release parameters as non-double-couple events had smaller stress drops and apparent stresses. The seismic efficiency of double-couple and non-double-couple events was limited to 0.9% and 0.05% respectively, with average values being 0.25% and 0.02%. A comparison of our results with those reported for a similar magnitude range in a hard-rock formation indicates considerably smaller estimates of stress drop and apparent stress in our case while seismic efficiencies are comparable.

Key words: Water injection, microseismic events, source parameters.

Introduction

It is widely accepted that large-scale fluid injections under high pressure can induce instabilities within a rock mass by reducing the effective normal stresses on pre-existing discontinuities, causing seismic activity similar to that for natural earthquakes. Since the observation of this phenomenon at Denver (HEALY *et al.*, 1968), Rangely (RALEIGH *et al.*, 1972) and Matsushiro (OHTAKE, 1974), similar observations at smaller scales have extended this conclusion to quite smaller microseismic events, particularly when clear evidence of a double-couple source was provided from fault-plane solutions (e.g., CASH *et al.*, 1983; TALEBI and CORNET, 1987). Other types of possible source mechanisms are related to the sudden extension of hydraulic fractures in tension and complex interactions of fluid and

[1] CANMET, 1079 Kelly Lake Rd., Sudbury, Ontario, Canada P3E 5P5. Fax: (705) 670-6556, E-mail: stalebi@.nrcan.gc.ca
[2] Imperial Oil Resources Ltd., 3535 Research Rd. NW, Calgary, Alberta, Canada T2L 2K8.

solid phases, considered to be at the origin of long-period events that show similar characteristics to volcanic tremors (BAME and FEHLER, 1986).

The literature on source parameters of seismic events induced by fluid injections, mainly concerns plutonic hard rock formations while few reports are available of similar attempts regarding sedimentary formations. The motivation behind the present study was to examine seismic source parameters of microseismic events ($M < -1$) recorded during water injection operations in a shale formation, and to compare the results to those of similar attempts in hard rock environment. The project was successful in simulating the occurrence of a casing failure in Colorado shales and established the technique as a viable detection tool for such applications (TALEBI et al., 1998).

Microseismic Activity

The data analyzed in this paper were recorded in September and October 1995 during water injection operations in the Cold Lake oil field in Alberta (TALEBI et al., 1998). The target of these operations was the Colorado shale formation at a depth of 220 m in a location remote from oil-producing wells. During the first two stages of the experiment, low flow rates of up to 12 m³/hr were used and no significant activity was detected. High flow rates of up to 30 m³/hr, used during the two following stages of the experiment, caused several hundred microseismic events to be recorded. The data acquisition system was set to record any seismic activity barely exceeding the background noise level. Figure 1 shows horizontal and vertical projections of the hypocenters of 135 events that were located. The sensor array consisted of two triaxial geophones cemented 199 m and 249 m deep in one borehole and strings of five hydrophones in three adjacent wells. The down-hole dual-gain preamplifiers of the geophone sensors were set to 84 dB and 104 dB. The sampling rate was 7.5 kHz/channel and the anti-aliasing filters, with a slope of 72 dB/octave, were set to 2 kHz. Calibration of the data acquisition system showed a flat frequency response up to 1.5 kHz.

Continuous recording of the background noise was made at a regular interval during the experiment in order to detect any changes in its properties as a function of time. Analysis of such recordings revealed the presence of low-frequency emissions with durations much longer than microseismic events (Fig. 2). The dominant frequency of these emissions was about 17 Hz (period of about 59 milliseconds) regardless of the sensor type or location. Rotating the geophone signals put all the energy on X components with a polarization direction pointing, invariably, to the injection well around the injection depth of 220 m. This observation indicates that analysis of such emissions, if recorded, can constitute a tool for detecting the presence of hydraulic fractures. Figure 3 displays an example of the raw spectra of such signals where a peaked structure is observed. The frequency peaks were remarkably similar for all the sensors and channels of recording. This phenomenon cannot be attributed to the recording instrumentation. The geophone sensors were cemented in place and the data acquisition system was calibrated before the experiment. Moreover, no such peaks were observed for calibration shots or the impulsive microseismic events. The observation clearly points to the presence of a resonance phenomenon within the seismically active area.

(a)

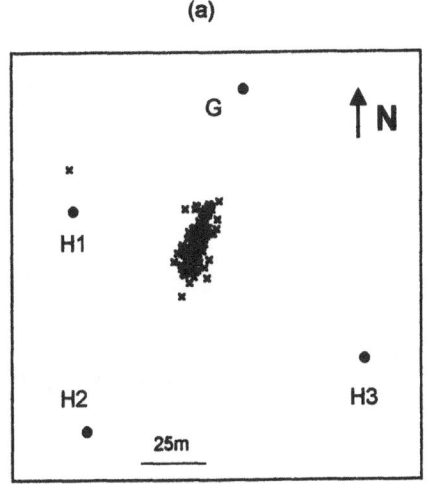

(b)

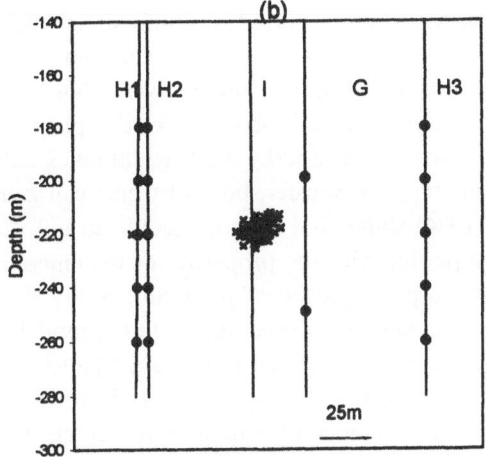

Figure 1

Plan view (a) and section along E-W (b) of event locations. Two triaxial geophones in borehole *G* and 15 hydrophones in boreholes *H*1, *H*2, and *H*3 were used to detect microseismicity associated with injection in the central borehole.

(a)

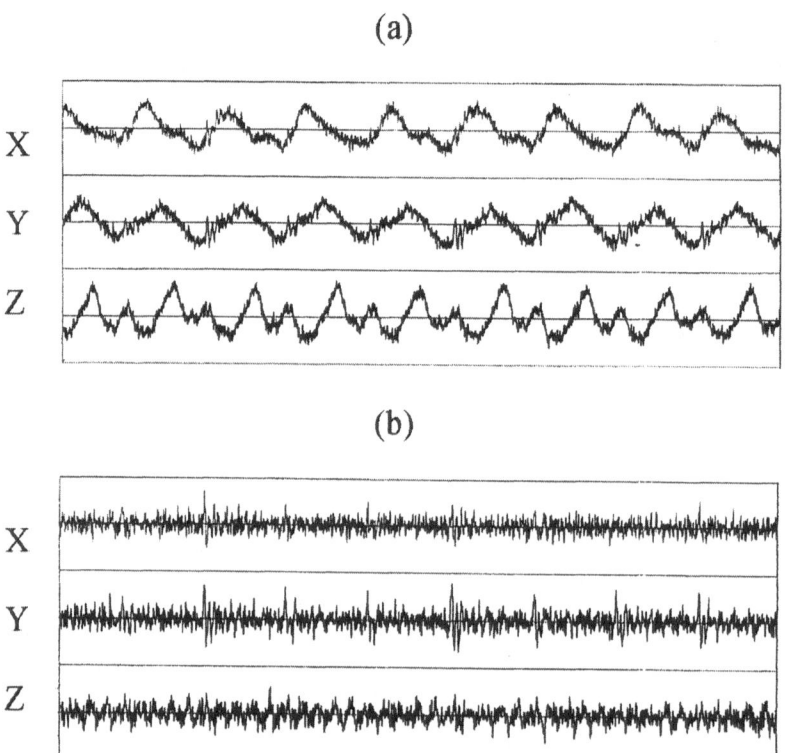

(b)

Figure 2

Example of signals of continuous emissions observed on the top geophone ($G1$): (a) raw signals, (b) high-pass filtered signals. The total time windows is 0.52 s.

Similar observations have been reported by authors in similar applications in oil fields (e.g., DOBECKI, 1989; FIX *et al.*, 1991). The similarities between events induced by fluid injections and volcanic tremors were stated by AKI *et al.* (1977) who were the first to attempt to model the source of this phenomenon. The sources considered by these authors included jerky incremental crack extension and sudden channel openings during magma transfer, both of which can generate tremors with peaked spectra. AKI (1984) states that tremor spectra do not depend on radiation direction, which is compatible with the properties of the emissions observed in the present case, and that the participation of the fluid activity should be included in crack models in order to explain the observations. Other models of volcanic tremor consider the transition of gas flow (STEINBERG and STEINBERG, 1975), unsteady fluid flow in conduits (FERRICK *et al.*, 1982), acoustic resonance of a fluid-filled pipe (CHOUET, 1985) and resonance of a fluid-driven crack (CHOUET, 1988). The polarized emissions in the present case became noticeable well into the injection operations. Similar to the observations in a previous experiment (TALEBI *et al.*, 1991), the characteristics of these emissions were compatible with resonances associated with water flow in a hydraulic fracture close to the injection point.

In the remainder of this paper, we focus our attention on the impulsive microseismic events (Fig. 1). The majority of located events showed clear *P* and *S* phases on the two geophones (Fig. 4). A collection of 12 events was selected for an in-depth analysis of their source parameters. Events 1–7 were recorded a few hours after the start of the first high flow-rate injection test while events 8–12 were recorded well into the second high flow-rate injection test. The intent was to detect any fundamental differences between the two sets as the maximum signal amplitudes observed had shown some increase with time during the experiment. Some of the selected events are somewhat stronger than the average recorded event however the subset is considered to be reasonably representative of the whole data set.

Methodology

The far-field displacement due to any reasonable kinematic model of an earthquake is expected to have a spectrum with a constant value at low frequencies and proportional to a negative power of frequency at high frequencies (AKI and RICHARDS, 1980). Three parameters which characterize such spectra are the flat

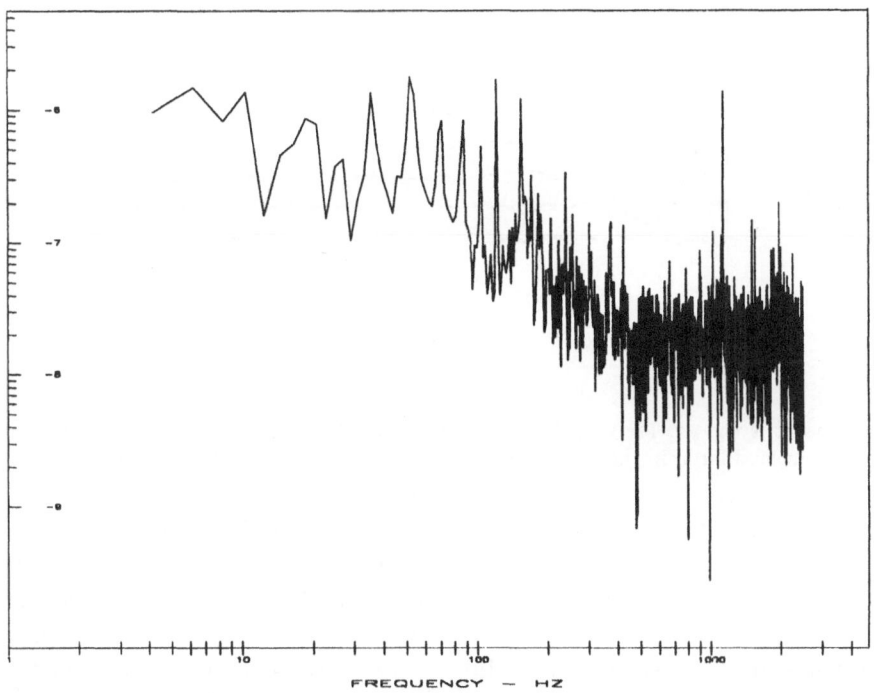

Figure 3
Example of raw spectra of signals of continuous emissions on a hydrophone sensor (the vertical axis is arbitrary).

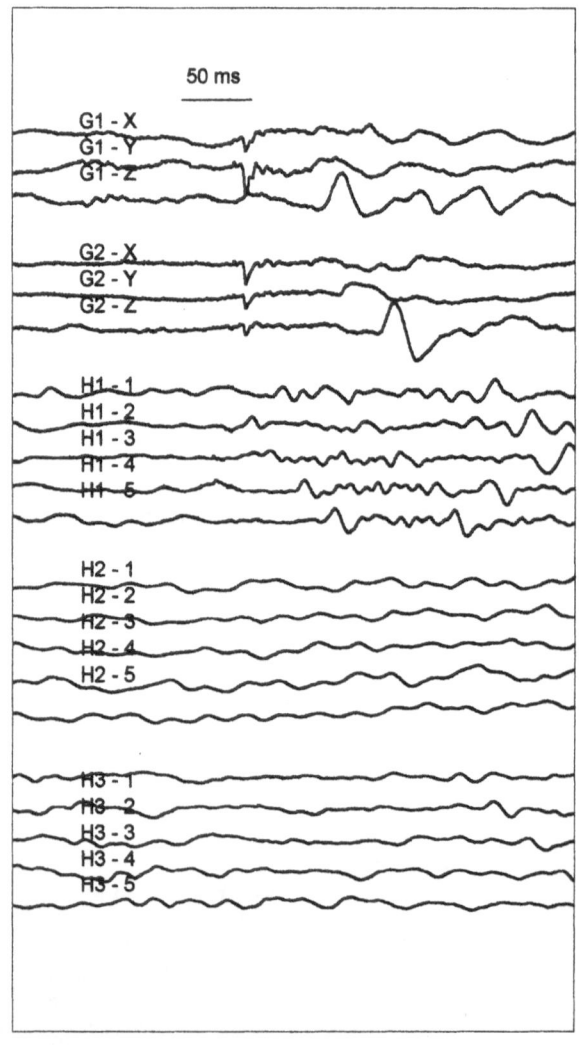

Figure 4

Raw signals of event #8. The vertical scale is identical for all geophone signals ($G1$ and $G2$), but is amplified for hydrophone signals by a factor of 7 for $H1$, 4 for $H2$ and 3 for $H3$.

low-frequency level Ω_0, the corner frequency corresponding to the intersection of the two trends f_0, and the power of the high-frequency asymptote γ. The first two parameters are generally used in the determination of seismic moment and source radius. Many kinematic models predict a value of -2 for the slope of the descending trend of spectra (e.g., BRUNE, 1970; SAVAGE, 1972; SATO and HI-RASAWA, 1973). The same is roughly the case for the dynamic model of MADAR-IAGA (1976) when rupture velocity is 0.9 times the shear-wave velocity (AKI and RICHARDS, 1980). According to RANDALL (1973), there are sound theoretical reasons for expecting the -2 model at high frequencies. HANKS (1979) derived the same conclusion based on a variety of seismic observations. The applicability of this model for events of moment magnitude in the range -3 to 3 has been reported by authors (e.g., GIBOWICZ et al., 1991; TRIFU et al., 1995). Based on the above, we consider that the following equation provides a reasonable fit to the far-field displacement spectrum of a shear type event:

$$|U(\omega)| = \frac{\Omega_0}{1 + \left(\dfrac{\omega}{\omega_0}\right)^2}, \tag{1}$$

where $\omega = 2\pi f$ is the angular frequency and $\omega_0 = 2\pi f_0$ is the angular corner frequency. The above equation was suggested by SNOKE (1987), based on the far-field pulse of the BRUNE's (1970) circular shear model. WALTER and BRUNE (1993) mention several successful applications of this equation to shear events and demonstrate that it is also consistent with kinematic models of circular tensile failure. Furthermore, such an approach would be consistent with the tensile model of SATO (1978) which predicts far-field displacement spectra with characteristics similar to those of shear models: i.e., a flat portion at low frequencies and a descending trend with a slope of -2 for frequencies higher than a corner frequency. Considering attenuation along the source-sensor path, we add an exponential term to Equation (1) to take this effect into account:

$$U(\omega) = \frac{\Omega_0 \cdot e^{-\omega R / 2CQ}}{1 + \left(\dfrac{\omega}{\omega_0}\right)^2}, \tag{2}$$

where R is the source-sensor distance, C is the velocity and Q is the quality factor of P or S wave. In the present case, R is in the range of 60 m to 85 m and average values of 1786 m/s and 643 m/s were accepted for P- and S-wave velocities based on the results of in situ measurements (TALEBI et al., 1998). The first step in signal processing consisted of rotating the recorded signals to obtain P, SH and SV records. By fitting Equation (2) to the rotated signals, three parameters were extracted from each trace: Ω_0, f_0 and Q. The procedure consisted of entering the initial estimates of these parameters by the operator and then using a least-squares technique to minimize an error function characterizing the difference between the observed and estimated spectra. Once the procedure had converged, the original spectra was divided by the exponential term in Equation (2) to obtain the corrected spectra. Figure 5 shows examples of the application of this technique.

The integral J_C of square of ground velocity of P- and S-wave signals was estimated from the corrected spectra using the relationship given by SNOKE (1987):

$$J_C = \tfrac{2}{3}[\Omega_0 \omega_1]^2 f_1 + 2 \int_{f_1}^{f_2} |\omega U(\omega)|^2 \, df + 2|\omega_2 U(\omega_2)|^2 f_2, \tag{3}$$

where f_1 and f_2 are the limits of the spectral bandwidth of the recording instrumentation. The results were used to recalculate corner frequencies according to:

$$f_0 = \left(\frac{J_C}{2\pi^3\Omega_0^2}\right)^{1/3}, \qquad (4)$$

where J_C for S waves is the sum for SH and SV components and Ω_0 is calculated vectorially from the low-frequency levels of SH and SV displacement spectra (GIBOWICZ and KIJKO, 1994). Visual inspection of the spectra revealed that this procedure produced more robust, although still comparable, results compared to those obtained in the previous step by fitting Equation (2) to individual traces. Seismic moments were then calculated from:

$$M_0 = \frac{4\pi\rho c^3 R\Omega_0}{F_C}, \qquad (5)$$

where ρ is the density of source material (2100 kg/m³) and F_C is the P- or S-wave radiation coefficient. RMS values of 0.52 and 0.63 were used for P and S waves (BOORE and BOATWRIGHT, 1984). Average values of seismic moment for P and S waves were accepted and applied to estimate moment magnitude M of the events (HANKS and KANAMORI, 1979):

$$M = 2/3 \log(M_0) - 6, \qquad (6)$$

where M_0 is seismic moment in N.m. The next step was to estimate seismic energies of P- or S-wave signals E_C:

$$E_C = 4\pi\rho c R^2 J_C \langle F_C^2\rangle/F_C^2, \qquad (7)$$

where $\langle\ \rangle$ represents the average value. Since focal mechanisms were not available in the present case, we assume that individual radiation coefficients are similar to their average values. P- and S-wave energies for each event were estimated by averaging the results of the two geophones for each wave type.

It is generally accepted that the dimensions of the failure area depend inversely on the corner frequency and that the radius of a circular source can be estimated from:

$$r_0 = \frac{kC}{2\pi f_0}, \qquad (8)$$

where k is a constant depending on the source model. There has been debate about the appropriate value of this coefficient to be used (e.g., GIBOWICZ, 1984; McGARR, 1991). For example, if we consider only the S waves, BRUNE (1970), SATO and HIRASAWA (1973), and MADARIAGA (1976) provide, respectively, values of 2.34, 1.85 and 1.32. The latter value is valid for the case in which rupture velocity is 0.9 times shear wave velocity. Estimates of source radii were obtained from P- and S-wave corner frequencies utilizing the above three models (considering the extension of Brune's model to P waves by HANKS and WYSS, 1972), although finally MADARIAGA's (1976) model was accepted. Some authors have indicated that the results from this model better fit the observed underground damage in mines (GIBOWICZ, 1984; TRIFU et al., 1995). Our motivation in adopting this model stems from the fact that, contrary to the other two models, it provided comparable source radii from P- and S-wave corner frequencies in most cases. Source radii calculated from the model are obviously about half those calculated from BRUNE's (1970) model.

Stress drop, defined as the average difference between the initial and final stress levels over the fault plane, was estimated from:

$$\Delta\sigma = \frac{7}{16}\frac{M_0}{r_0^3}. \tag{9}$$

Apparent stress was calculated from the estimates of seismic moment and total seismic energy ($E = E_P + E_S$):

$$\sigma_{app} = \frac{\mu E}{M_0} = \eta\tau, \tag{10}$$

where μ is the dynamic shear modulus estimated at 0.87 GPa, η is seismic efficiency and τ is the average shear stress before and after failure.

Results

Processing of the data established that Equation (2) provided a satisfactory fit to both P- and S-wave displacement spectra (Fig. 5). The process converged in about two thirds of the cases and estimates of quality factor were obtained. The twelve events recorded on the two geophones provided a total of 72 seismic traces of P, SH and SV components (24 of each). The number of resonable estimates of Q was, respectively, 18, 10 and 19. Figure 6 shows Q_S (defined as $1/[1/Q_{SH} + 1/Q_{SV}]$) versus Q_P for the cases in which both values were available for the same event and the same sensor. The best fit to the data provides a ratio of $Q_P/Q_S = 2.5$. Table 1 presents the average results for individual events. Average values of $Q_P = 34$ and $Q_S = 15$ were obtained after excluding the results of the first event that were unusually high. These average values seem to be reasonable estimates for the shale formation studied here. On the one hand, they are compatible with the in situ measurements of MCDONAL et al. (1958) for a similar formation ($Q_P = 32$; $Q_S = 10$) and the results of JOHNSTON and TOKSÖZ (1980) for Colorado oil shales. On the other hand, they are in agreement with attenuation measurements in the same formation using signals from calibration shots (TALEBI et al., 1998). The spectra of the traces in which convergence was not satisfactory were corrected for attenuation using these average values before estimating source parameters.

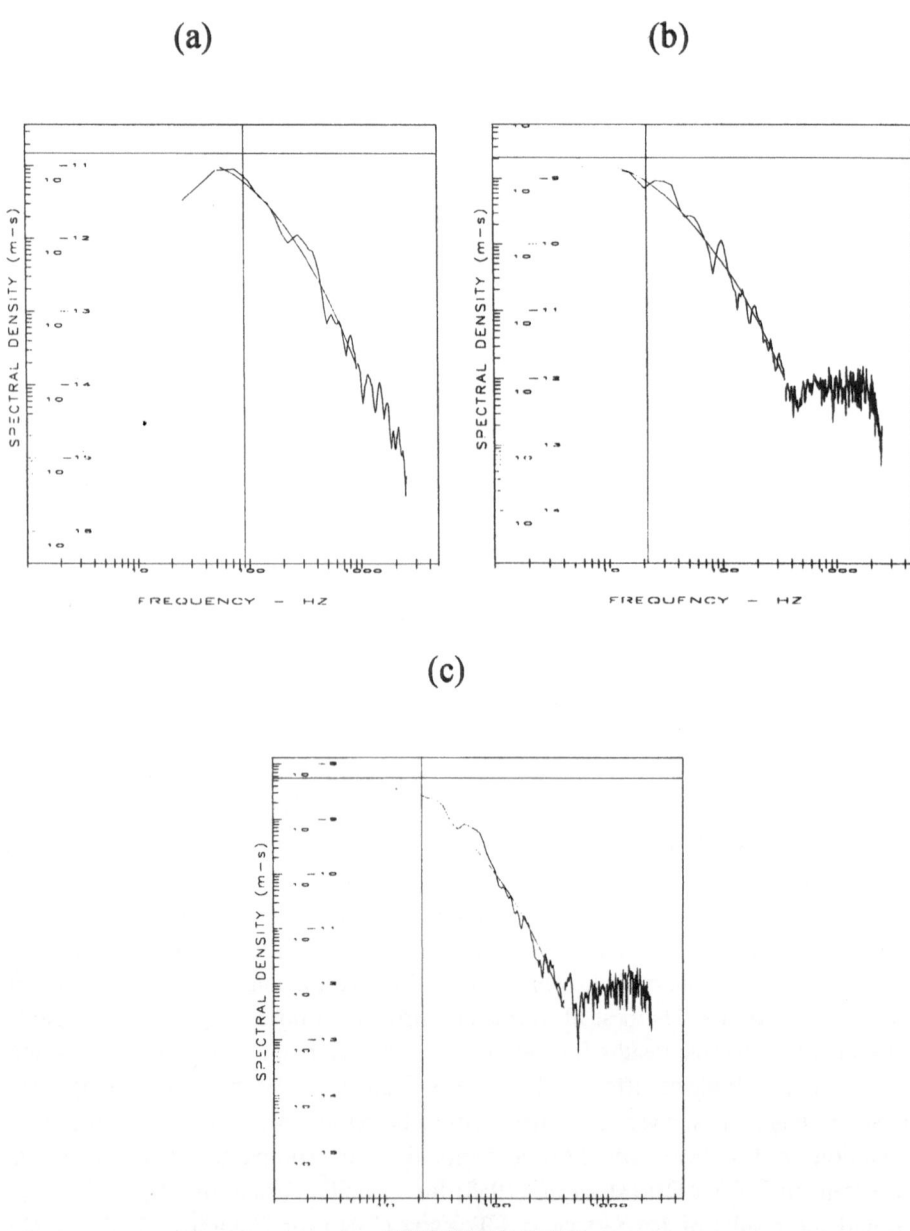

Figure 5

Typical examples of displacement spectra, corrected for attenuation, of *P* (a), *SH* (b) and *SV* (c) components. The best-fit curves to the spectra are also shown. The horizontal and vertical lines indicate, respectively, the flat low-frequency trend and the corner frequency.

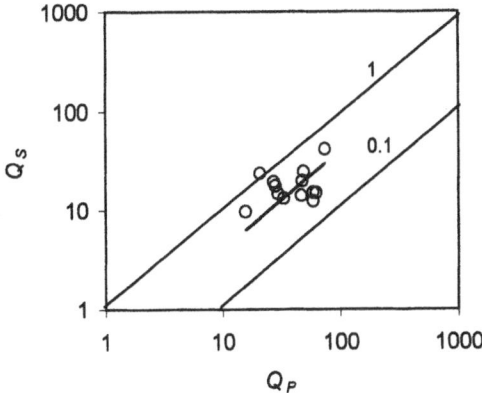

Figure 6

Q_S versus Q_P where both measurements were available on the same sensor for the same event. Straight lines show constant values of Q_S/Q_P and the best fit value of 0.4.

Table 1 summarizes the results of source parameter determinations. Moment magnitudes for the present data set are in the range of -3 to -1.1. We first consider the partition of energy between P and S waves and investigate the use of energy ratios in the detection of the non-double-couple component of failure. Table 1 indicates that for three events E_S/E_P is smaller than 10 (see also Fig. 7). This value is sometimes taken as the lower limit for what can be expected for pure shear failure. The first event in Table 1 was strong enough to be recorded on all available sensors. The first motion of P waves for this event was clearly compressional on all

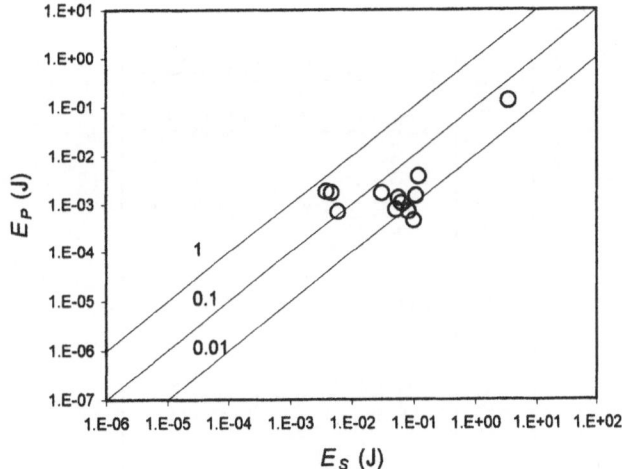

Figure 7

P-wave energy versus S-wave energy; straight lines show constant values of E_P/E_S.

Shahriar Talebi and Tom J. Boone

Table 1

Results of source parameter determinations. Events 1, 3, 5 and 6 have a non-double-couple component

Event	M	Q_P	Q_S	M_0 (Nm)	E (mJ)	E_P (mJ)	E_S (mJ)	E_S/E_P	r_0 (m)	$\Delta\sigma$ (kPa)	σ_{app} (kPa)	σ_{app}/τ
1	−1.1	80	45	2.1×10^7	3553.3	132.9	3420.4	25.7	10.5	7.8	0.15	5.0×10^{-4}
2	−3.0	27	19	3.1×10^4	32.2	1.7	30.5	18.3	0.7	35.5	0.90	3.0×10^{-3}
3	−2.6	48	20	1.3×10^5	6.4	1.7	4.6	2.7	2.0	7.2	0.04	1.3×10^{-4}
4	−2.8	16	10	7.3×10^4	124.4	3.8	120.7	32.1	0.9	39.1	1.48	5.0×10^{-3}
5	−2.5	28	—	1.9×10^5	5.7	1.8	3.8	2.1	3.7	1.7	0.03	1.0×10^{-4}
6	−2.2	49	25	5.6×10^5	6.6	0.7	5.9	8.6	4.6	2.6	0.01	3.3×10^{-5}
7	−3.0	47	14	3.7×10^4	111.2	1.5	109.8	73.7	0.6	71.7	2.64	8.8×10^{-3}
8	−2.4	59	12	2.5×10^5	52.5	0.8	51.7	68.1	2.9	4.4	0.19	6.3×10^{-4}
9	−2.3	59	15	4.2×10^5	83.5	0.7	82.8	117.4	3.1	6.1	0.17	5.7×10^{-4}
10	−2.2	30	15	5.3×10^5	101.8	0.4	101.4	226.9	3.1	7.8	0.17	5.7×10^{-4}
11	−2.4	28	17	2.6×10^5	64.9	1.1	63.8	60.5	1.6	28.9	0.22	7.3×10^{-4}
12	−2.3	33	13	3.4×10^5	58.8	1.3	57.4	42.8	3.0	5.7	0.15	5.0×10^{-4}

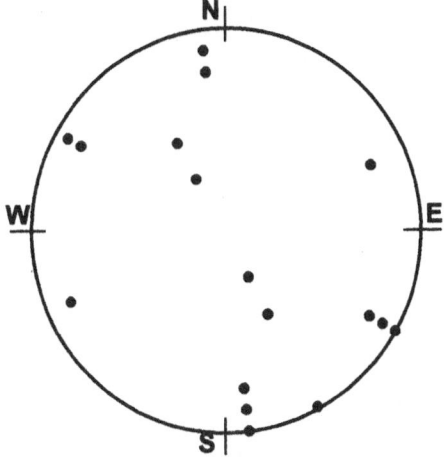

Figure 8
Equal-area projection of first motion of *P* waves for the first event. A compressional motion (solid circles) was clearly observed at all the sensors.

the sensors (Fig. 8). This strongly suggests a dominant component of tensile failure at the source of this event. The possibility of some shear component cannot be discarded, however, since E_S/E_P for this event is 26. We consider that the four events mentioned above (numbers 1, 3, 5 and 6 in Table 1) have some non-double-couple component at their source. All of these four events were recorded a few hours after the start of the first injection operation. This observation, which constitutes the main difference between the two sets of analyzed events, is likely to be related to the propagation of hydraulic fractures according to a tensile mode of failure early in the injection operation.

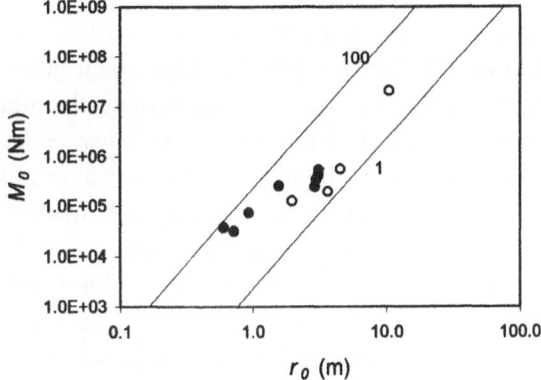

Figure 9
Seismic moment versus source radius. Open circles indicate the events with a non-double-couple component. Straight lines show constant values of stress drop in kPa.

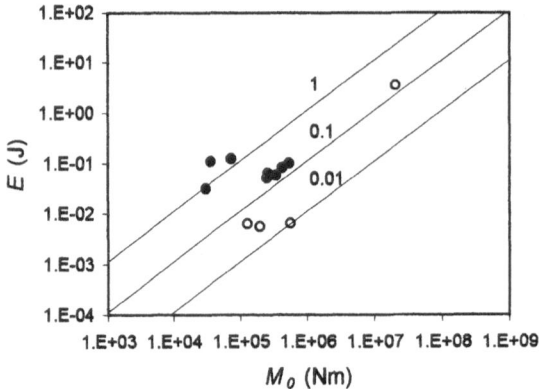

Figure 10
Total seismic energy E versus seismic moment M_0. Open circles indicate the non-double-couple events.
Straight lines show constant values of apparent stress in kPa.

Figure 9 shows seismic moment versus source radius, where straight lines correspond to constant stress drops of 1 kPa and 100 kPa (1 kPa = 0.01 bar). Note that our stress drops are two orders of magnitude lower than the range of 1–100 bars often reported for earthquakes. This is consistent with observations of smaller stress drops for other very small events in the literature. Evidence of small stress drops (< 1 bar) has been reported for hydraulic fracturing events with comparable source radii (FEHLER and PHILLIPS, 1991) and small earthquakes (e.g., FRANKEL, 1981; FLETCHER et al., 1986). The non-double-couple events tend to have smaller stress drops and apparent stresses (Figs. 9 and 10). This is clearly related to the lower E_S/E_P ratios obtained for these events and corroborates the observations of GIBOWICZ et al. (1990, 1991).

The different components of the stress field and their implications on the orientation of hydraulic fractures in the area of the Cold Lake oil field have been discussed by DUSSEAULT (1977). At depths less than about 400 m, the minimum principal stress component is in the vertical direction and the other two principal components are horizontal. At a depth of 220 m where microseismicity was observed in the present experiment, the results of the previous author, which have been confirmed by recent stress measurements by hydraulic fracturing, indicate that the maximum effective principal stress is in the range of 3.8–4.4 MPa and the minimum effective principal stress is about 2.4 MPa. Therefore, the maximum available shear stress at this depth is in the range of 0.7–1.0 MPa. The increase in pore pressure due to water injection is expected to reduce the normal effective stresses, but should not affect shear stresses on fracture planes. The amplitude of shear stresses acting on fracture planes prior to failure obviously depends on the orientation of such fractures but cannot exceed 1 MPa. We assume that these stresses are within the range of 0.2–1.0 MPa. Accordingly, assuming total stress release following a failure episode, the average shear stress before and after failure (τ) should be in range of 0.1–0.5 MPa. We consider that the average value of 0.3 MPa for this parameter should provide reasonable estimates within a factor of 2. Seismic efficiency $(\eta = \sigma_{app}/\tau)$ for each event was estimated based on this value and reported in the last column of Table 1. The upper bound of seismic efficiency of double-couple events is 0.9% with an average value of 0.25%. The non-double-couple events have lower seismic efficiencies with upper bound and average values of 0.05% and 0.02%, respectively.

The data set analyzed by GIBOWICZ et al. (1991) was recorded in a hard-rock underground research laboratory and covered a moment magnitude range of −3.6 to −1.9 which is close to the range of our events (− 3.0 to − 1.1). We consider their results as a reference and attempt to detect any fundamental differences with our results. MCGARR (1994) estimates seismic efficiencies for the events analyzed by GIBOWICZ et al. (1991) to be in the range of 0.0003 to 0.051. Since apparent stresses varied mostly around a value of about two bars in their results, a reasonable value of seismic efficiency for the majority of events would be about 0.5%. This value is comparable with the lower bound of estimates of seismic efficiency for Denver earthquakes (MCGARR, 1976) and the upper bound of estimates for mine-induced events (MCGARR et al., 1979). It is also in reasonable agreement with the results of the present study for double-couple events. The main difference between our results and those of GIBOWICZ et al. (1991) seems to be in the estimates of stress release parameters, as our stress drops and apparent stresses are about two orders of magnitude smaller.

Conclusion

The observed microseismic activity included two types of events: several hundred impulsive events with clear P- and S-wave arrivals, and continuous emissions with no clear arrivals. The second type of activity showed a peaked spectra on all channels of recording. The polarization direction of these emissions points, invariably, to the injection point, compatible with resonances associated with water flow within a hydraulic fracture.

For a representative collection of impulsive events, an ω^{-2} model provided satisfactory fits to displacement spectra corrected for attenuation and average quality factors of 34 and 15 for P and S waves. P-wave first motion analysis and E_S/E_P ratios indicated the existence of a non-double-couple component in some events, particularly early in the experiment. Seismic efficiency of double-couple and non-double-couple events was limited to 0.9% and 0.05%, with average values of 0.25% and 0.02%, respectively. Also, a clear difference was observed between estimates of stress release parameters for the two types of events as non-double-couple events had smaller stress drops and apparent stresses. Comparison of the present source parameters with those reported for a similar magnitude range in a hard-rock formation indicates that our estimates of stress drop and apparent stress are two orders of magnitude smaller, although seismic efficiencies are comparable.

Acknowledgments

The authors would like to thank Imperial Oil Limited, Resources Division, and CANMET for their support of this project and their permission to publish the data. Data analysis benefitted from the assistance of S. Ogunye and P. Rochon at CANMET. We are grateful to A. McGarr and S. J. Gibowicz for their careful reviews which helped to improve the quality of this paper.

REFERENCES

AKI, K. (1984), *Evidence for Magma Intrusion During the Mammoth Lakes Earthquakes of May 1980 and Implications for the Absence of Volcanic (Harmonic) Tremor*, J. Geophys. Res. *89*, 7689–7696.

AKI, K., and RICHARDS, P. G., *Quantitative Seismology* (W. H. Freeman, San Francisco 1980).

AKI, K., FEHLER, M., and DAS, S. (1977), *Souce Mechanism of Volcanic Tremor: Fluid-driven Crack Models and their Application to the 1963 Kilauea Eruptions*, J. Volcan. Geoth. Res. *2*, 259–287.

BAME, D., and FEHLER, M. (1986), *Observations of Long-period Earthquakes Accompanying Hydraulic Fracturing*, Geophys. Res. Lett. *13*, 149–152.

BOORE, D. M., and BOATWRIGHT, J. (1984), *Average Body-wave Radiation Coefficients*, Bull. Seismol. Soc. Am. *74*, 1615–1621.

BRUNE J. N. (1970), *Tectonic Stress and the Spectra of Seismic Shear Waves from Earthquakes*, J. Geophys. Res. *75*, 4997–5009.

CASH, D., HOMUTH, E. F., KEPPLER, H., PEARSON, C., and SASAKI, S. (1983), *Fault Plane Solutions for Microearthquakes Induced at the Fenton Hill Hot Dry Rock Geothermal Site: Implication for the State of Stress near a Quarternary Volcanic Center*, Geophys. Res. Lett. *10*, 1141–1144.

CHOUET, B. (1985), *Excitation of a Buried Magmatic Pipe: A Seismic Source Model for Volcanic Tremor*, J. Geophys. Res. *90*, 1881–1893.

CHOUET, B. (1988), *Resonance of a Fluid-driven Crack: Radiation Properties and Implications for the Source of Long-period Events and Harmonic Tremor*, J. Geophys. Res. *93*, 4375–4400.

DOBECKI, T. L., *Acoustic Emissions from Hydraulic Fractures: Long-term Observations*. In *AE/MS in Geologic Structures and Materials* (Hardy Jr., H. R., ed.) (Trans. Tech. Publications, Germany 1989) pp. 403–415.

DUSSEAULT, M. B. (1977), *Stress State and Hydraulic Fracturing in the Athabasca Oil Sands*, J. Canadian Petroleum Technology *16*, 19–27.

FEHLER, M., and PHILLIPS, W. S. (1991), *Simultaneous Inversion for Q and Source Parameters of Microearthquakes Accompanying Hydraulic Fracturing in Granitic Rock*, Bull. Seismol. Soc. Am. *81*, 553–575.

FERRICK, M. G., QAMAR, A., and ST. LAWRENCE, W. F. (1982), *Source Mechanism of Volcanic Tremor*, J. Geophys. Res. *87*, 8675–8683.

FIX, J. E., FRANTZ, J. H., and LANCASTER, D. E. (1991), *Applications of microseismic technology in a Devonian shale well in the Appalachian Basin*. In Proc. SPE Eastern Regional Meeting, Lexington, Kentucky, October 22–25, 1991, 109–122.

FLETCHER, J. B., HARR, L. C., VERNON, F. L., BRUNE, J. N., HANKS, T. C., and BERGER, J., *The effects of attenuation on the scaling of source parameters for earthquakes at Anza, California*. In *Earthquake Source Mechanics* (S. Das, J. Boatwright, and C. Scholz, eds.) (AGU, Washington D.C. 1986) pp. 331–338.

FRANKEL, A. (1981), *Source Parameters and Scaling Relationships of Small Earthquakes in the North-eastern Caribbean*, Bull. Seismol. Soc. Am., *71*, 1171–1190.

GIBOWICZ, S. J., *The mechanism of large mining tremors in Poland*. In *Proc. 1st Int. Cong. On Rockbursts and Seismicity in Mines* (Gay, N. C., and Wainwright, E. H., eds.) (South African Institute of Mining and Metallurgy, Johannesburg 1984) pp. 17–28.

GIBOWICZ, S. J., HARJES, H.-P., and SCHÄFFER, M. (1990), *Source Parameters of Seismic Events at Heinrich Robert Mine, Rhur Basin, Federal Republic of Germany: Evidence for Non-double-couple Events*, Bull. Seismol. Soc. Am. *80*, 80–109.

GIBOWICZ, S. J., and KIJKO, A., *An Introduction to Mining Seismology* (Academic Press, San Diego 1994).

GIBOWICZ, S. J., YOUNG, R. P., TALEBI, S., and RAWLENCE, D. J. (1991), *Source Parameters of Seismic Events at the Underground Research Laboratory in Manitoba, Canada: Scaling Relations for Events with Moment Magnitude Smaller than −2*, Bull. Seismol. Soc. Am. *81*, 1157–1182.

HANKS, T. C. (1979), *b values and $\omega^{-\gamma}$ Seismic Source Models: Implications for Tectonic Stress Variations along Active Crustal Fault Zones and the Estimation of High-frequency Strong Ground Motion*, J. Geophys. Res. *84*, 2235–2242.

HANKS, T. C., and KANAMORI, H. (1979), *A Moment Magnitude Scale*, J. Geophys. Res. *84*, 2348–2350.

HANKS, T. C., and WYSS, M. (1972), *The Use of Body-wave Spectra in the Determination of Seismic-source Parameters*, Bull. Seismol. Soc. Am. *62*, 561–589.

HEALY, J. H., RUBEY, W. W., GRIGGS, D. T., and RALEIGH, C. B. (1968), *The Denver Earthquakes*, Science *161*, 1301–1310.

JOHNSTON, D. H., and TOKSÖZ, M. N. (1980), *Ultrasonic P- and S-wave attenuation in Dry and Saturated Rocks under Pressure*, J. Geophys. Res. *85*, 925–936.

MADARIAGA, R. (1976), *Dynamics of an Expanding Circular Fault*, Bull Seismol. Soc. Am. *66*, 639–666.

McDONAL, F. J., ANGONA, F. A., MILLS, R. L., SENGBUSH, R. L., VAN NOSTRAND, R. G., and WHITE, J. E. (1958), *Attenuation of Shear and Compressional Waves in Pierre Shale*, Geophysics *23*, 421–439.

McGARR, A. (1976), *Seismic Moments and Volume Changes*, J. Geophys. Res. *81*, 1487–1494.

McGARR, A. (1991), *Observations Constraining Near-source Ground Motion Estimated from Locally Recorded Seismograms*, J. Geophys. Res. *96*, 16495–16508.

McGARR, A. (1994), *Some Comparisons between Mining-induced and Laboratory Earthquakes*, Pure appl. geophys. *142*, 467–489.

McGARR, A., SPOTTISWOODE, S. M., GAY, N. C., and ORTLEPP, W. D. (1979), *Observations Relevant to Seismic Driving Stress, Stress Drop and Efficiency*, J. Geophys. Res. *84*, 2251–2261.

OHTAKE, M. (1974), *Seismic Activity Induced by Water Injection at Matsushiro, Japan*, J. Phys. Earth *22*, 163–176.

RALEIGH, C. B., HEALY, J. H., and BREDEHOEFT, J. D., *Faulting and crustal stress at Rangely, Colorado*. In *Flow and Fracture of Rocks* (American Geophysical Union, Washington, D.C. 1972) pp. 275–284.

RANDALL, M. J. (1973), *The Spectral Theory of Seismic Sources*, Bull. Seismol. Soc. Am. *63*, 1133–1144.

SATO, T. (1978), *A Note on Body-wave Radiation from Expanding Tension Cracks*, Sci. Rep. Tohoku Univ., Ser. 5, Geophysics *25*, 1–10.

SATO, T., and HIRASAWA, T. (1973), *Body-wave Spectra from Propagating Shear Cracks*, J. Phys. Earth *21*, 415–431.

SAVAGE, J. C. (1972), *Relation of Corner Frequency to Fault Dimensions*, J. Geophys. Res. *77*, 3788–3795.

SNOKE, J. A. (1987), *Stable Determination of (Brune) Stress Drop*, Bull. Seismol. Soc. Am. *77*, 530–538.

STEINBERG, G., and STEINBERG, A. S. (1975), *On Possible Causes of Volcanic Tremor*, J. Geophys. Res. *80*, 1600–1604.

TALEBI, S., and CORNET, F. H. (1987), *Analysis of the Microseismicity Induced by a Fluid Injection in a Granitic Rock Mass*, Geophys. Res. Lett. *14*, 227–230.

TALEBI, S., BOONE, T. J., and EASTWOOD, J. E. (1998), *Injection-induced Microseismicity in Colorado Shales*, Pure appl. geophys. *153*, 95–111.

TALEBI, S., VANDAMME, L., McGAUGHEY, W. J., and YOUNG, R. P., *Microseismic mapping of a hydraulic fracture*. In *Rock Mechanics as a Multidisciplinary Science* (Roegiers, J. C., ed.) (Balkema, Rotterdam 1991) pp. 461–470.

TRIFU, C-I., URBANCIC, T. I., and YOUNG, R. P. (1995), *Source Parameters of Mining-induced Seismic Events: An Evaluation of Homogeneous and Inhomogeneous Faulting Models for Assessing Damage Potential*, Pure appl. geophys. *145*, 3–27.

WALTER, W. R., and BRUNE, J. N. (1993), *Spectra of Seismic Radiation from a Tensile Crack*, J. Geophys. Res. *98*, 4449–4459.

(Received June 19, 1998, revised/accepted August 10, 1998)

Seismicity Triggered by Resevoirs and Aquifers

Pure appl. geophys. 153 (1998) 133–149
0033–4553/98/010133–17 $ 1.50 + 0.20/0

❙Pure and Applied Geophysics

Reservoir-induced Seismicity in China

LINYUE CHEN[1] and PRADEEP TALWANI[1]

Abstract—A review of case histories of reservoir-induced seismicity (RIS) in China shows that it mainly occurs in granitic and karst terranes. Seismicity in granitic terranes is mainly associated with pore pressure diffusion whereas in karst terranes the chemical effect of water appears to play a major role in triggering RIS. In view of the characteristic features of RIS in China, we can expect moderate earthquakes to be induced by the construction of the Three Gorges Project on the Yangtze River.

Key words: Reservoir-induced seismicity in China, mechanism of reservoir-induced seismicity.

Introduction

Since the earliest, and one of the most destructive cases of reservoir-induced seismicity (RIS), the M_s 6.1 earthquake in Xinfengjiang in 1962, there have been 18 other cases of RIS in China. These have ranged in magnitude between M_s 4.8 and 2.2, and have occurred in different geologic zones and have been associated with a large range of impoundment histories and water levels. The RIS at these reservoirs has been the subject of non-uniform studies, ranging from detailed studies at Xinfengjiang and Danjiangkou Reservoirs, to very few studies at Shenjiaxia and Shuikou. Very few details are available in English and most accounts are only available in various technical reports, journals and books in Chinese.

Two large hydroelectric projects, the Three Gorges Project on the Yangtze River and the Xiaolangdi Project on the Yellow River (Fig. 1) are currently under construction. When completed, they will be among the largest in the world and in view of the incidence of RIS in China, it is important to understand and assess their seismic potential. In order to do so, it is necessary to understand the nature of the RIS that has been observed to date. Towards that end, in this paper we review the lithology of hypocentral areas where RIS has been observed, their filling history and reservoir characteristics.

---

[1] Department of Geological Sciences, University of South Carolina, Columbia, South Carolina, U.S.A.

Mechanism of RIS

Recently TALWANI (1997) reviewed our current understanding of the nature of RIS. The emphasis was on the physical effect of impoundment on triggering RIS. The strength changes (S) are governed by the Coulomb criterion, $S = S_0 + \mu(\sigma_n - p)$ where S_0 is cohesion, μ the coefficient of friction, σ_n the normal stress and p the pore pressure. Increase in pore pressure, or a decrease in the coefficient of friction and cohesion results in weakening the rocks and leading them to failure.

The elastic response to impoundment is instantaneous and manifests itself by an increase in the normal stress. It results in stabilizing the regions below the deepest part of the reservoir, resulting in little or no activity there. The impounding of a reservoir also results in an instantaneous increase in pore pressure, due to Skempton's effect. This effect is usually short-lived. As the pore pressure diffuses away, there is a cessation in seismicity. The delayed effect of pore pressure diffusion from the reservoir to hypocentral depths is most widely observed. Pore pressures increase away from the reservoir (thus weakening the rocks), resulting in a delay between the filling and the onset of seismicity. The seismicity occurs on the periphery of the reservoir and is usually associated with outward migration of epicenters along faults. Thus, delayed seismicity is most commonly observed.

In recent years we have come to recognize that the impoundment of a reservoir can reduce the strength of rocks by reducing the coefficient of friction, μ, or by

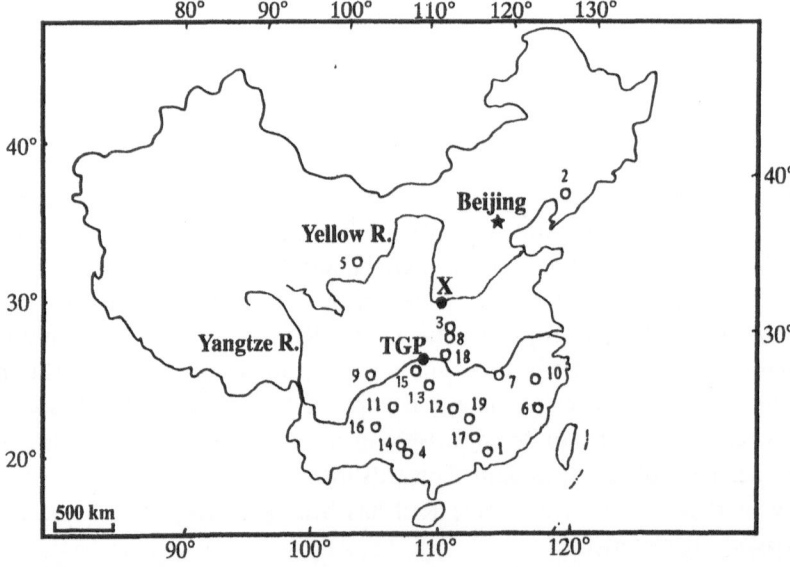

Figure 1
Map showing the locations of RIS cases in China and the Three Gorges and Xiaolangdi Projects (modified from HU *et al.*, 1996). TGP, Three Gorges Project; X, Xiaolangdi Project.

reducing the cohesive strength of rocks, S_0. Introduction of water or an increase in pore pressure in clayey gouge leads to a lowering of μ (TALWANI and ACREE, 1984/85). The cohesive strength decreases due to stress corrosion and dissolution on carbonate rocks. Chemical dissolution for carbonates occurs in the crust according to the following chemical reaction (FYFE *et al.*, 1978) $CaCO_3 + CO_2 + H_2O = Ca(HCO_3)_2$. When water diffuses to greater depths in carbonate rocks, and there is an increase in the CO_2 content, dissolution occurs. Dissolution results in a decrease in cohesion (and sometimes the coefficient of friction). Failure also occurs because of karst cavity collapse, which results from dissolution of the carbonate rocks and the added load of the reservoir.

In China we find evidence that all these mechanisms play an important role in RIS. Next we review our knowledge of RIS in China.

RIS in China

Figure 1 illustrates the locations of RIS in China and the projects under construction; the Xiaolangdi and Three Gorges Projects. The height of the dam, reservoir volume, dates of impoundment, initial seismicity, and the magnitude and date of the largest event are given in Table 1. Most of the reservoirs are in granitic or in carbonate rocks (mostly karsts). The hypocentral lithology, presence of nearby active faults, and the depths of the main shock and depth range of predominant seismicity are given in Table 2.

Following the RIS near Xinfengjiang Reservoir (No. 1), there have been 18 more cases of RIS in China. Except for the Shenwo Reservoir (No. 2) in Liaoning Province and the Shengjiaxia Reservoir (No. 5) in Qinghai Province, all other cases of RIS occurred in South China (Fig. 1).

There are 348 reservoirs in China with a volume of 0.1 km^3 or greater, of these 15 (4.4%) are associated with RIS. Of these 348 "large" reservoirs, 82 have active faults within 3 km of the reservoir. Of these 82, five reservoirs (6.1%) (Table 2) have RIS associated with them (CHEN, 1995). Thus the fraction of "large" reservoirs with active faults and RIS is similar to the fraction of "large" reservoirs with RIS. Consequently the presence of nearby active faults is not a determining factor in anticipating RIS.

Earthquake Locations and Lithology

Two lithologies are prevalent in the hypocentral areas of the 19 cases listed in Tables 1 and 2. Except for Danjiangkou (No. 3) the rocks at hypocentral depths were the same as those at the epicenters. Correspondingly in Table 2, the lithologies are for hypocentral depths. The hypocentral areas lie in granitic (4) and carbonate

Table 1

General characteristics of RIS in China

No.	Reservoir name	Dam height (m)	Volume (km^3)	Initial filling	Initial activity	Largest M_s (Io*)	Earthquake date	Reference no.
1	Xinfengjiang	105	11.5	Oct. 20 1959	Nov. 1959	6.1 (VIII)	Mar. 19 1962	(1)
2	Shenwo	50.3	0.54	Nov. 1 1972	Feb. 1973	4.8 (VI)	Dec. 22 1974	(2)
3	Danjiangkou	97	19	Nov. 1967	Jan. 1970	4.7 (VII)	Nov. 29 1973	(3)
4	Dahua	74.5	0.42	May 27 1982	Jun. 4 1982	4.5 (VII)	Feb. 10 1993	(4)
5	Shengjiaxia	35	0.004	Oct. 1980	Nov. 1981	3.6 (VI)	Mar. 7 1984	(5)
6	Shuikou	101	2.35	May 1993	Jul. 1993	3.2 (VI)	Jan. 12 1994	(5)
7	Zhelin	62	7.17	Jan. 31 1971	Feb. 1971	3.2 (V)	Oct. 14 1972	(6)
8	Qianjin	50	0.02	May 1970	Oct. 20 1971	3.0 (VI)	Oct. 20 1971	(7)
9	Tongjiezi	74	0.03	Apr. 5 1992	Apr. 6 1992	2.9 (V)	Jul. 17 1992	(8)
10	Hunanzhen	129	2.06	Jan. 12 1979	Jun. 28 1979	2.8 (V)	Oct. 7 1979	(9)
11	Wujiangdu	165	2.14	Nov. 20 1979	Mar. 1980	2.8 (V)	Mar. 7 1985	(5)
12	Nanchong	45	0.015	1969	1969	2.8 (VI)	Jul. 25 1974	(10)
13	Huangshi	40.5	0.61	Jan. 1970	May 1973	2.8 (V)	Sept. 21 1974	(11)
14	Yantan	110	2.43	Mar. 19 1992	Mar. 29 1992	$M_L 3.5$ (V)	Jun. 21 1994	(4)
15	Geheyan	151	3.4	Apr. 1993	Apr. 1993	2.6 (V)	May 30 1993	(9)
16	Lubuge	103	0.11	Nov. 21 1988	Nov. 24 1988	$M_L 3.4$ (VI)	Dec. 17 1988	(12)
17	Nanshui	81.5	1.22	Feb. 1969	Jan. 1970	2.3 (V)	Feb. 26 1970	(13)
18	Dengjiaqiao	12	0.0004	Dec. 1979	Aug. 1 1980	2.2 (V)	Oct. 30 1983	(14)
19	Dongjiang	157	8.12	Aug. 2 1986	Nov. 1986	$M_L 3.2$ (>V)	Jul. 20 1991	(15)

*Io: Epicentral Intensity: (1) DING et al. (1987), (2) ZHONG et al. (1981), (3) GAO and YING (1980), (4) GUANG (1995), (5) HU et al. (1996), (6) HUANG and KONG (1984), (7) GAO et al. (1984), (8) GUO (1994), (9) HU et al. (1986), (10) HU and CHEN (1979), (11) KONG (1984), (12) JIANG and WEI (1995), (13) XIAO and PAN (1984), (14) LIU and LI (1981), (15) HU et al. (1995).

(15) rocks (Table 2). We have adequate data for only two of the four locations in granitic rocks, Xinfengjiang and Hunanzhen Reservoirs (Nos. 1 and 10). RIS at Xinfengjiang Reservoir has been the subject of several detailed studies and we will

address it in a later section. Of the 15 cases in carbonate rocks we have adequate data for 14 reservoirs. In 13 of these the hypocenters were in carbonate rocks.

As the nearest station was about 60 km distant, the location of the Shenwo earthquake (No. 2) is not well constrained. The macroscopic epicenter is about 7 km from the instrumentally determined epicenter. They both lie in middle Ordovician limestone (ZHONG et al., 1981). The instrumentally determined depth, ~6 km and the depth of the predominant seismicity (4 to 8 km) all lie in Proterozoic dolomites.

A nine station seismic network was installed at Danjiangkou Reservoir (No. 3) after the M 4.7 earthquake in January 1970. The depth of the subsequent seismicity is accurate to about ±1 km and lay between 2 and 5 km in Precambrian dolomites. The depth of the main shock (5 to 9 km) is not well constrained and the main shock may have occurred in the underlying Proterozoic metamorphic rocks. The columnar stratigraphy is from YANG et al. (1986).

A five station seismic network was installed before the impoundment of Dahua (No. 4) and the nearby Yantan (No. 14) reservoirs. The hypocentral depths are

Table 2

Geological settings and focal depths of RIS in China

No.	Reservoir name	Hypocenter lithology	Active faults	Earthquake depths (km)			Reference no.
				predominant	main shock	accuracy	
1	Xinfengjiang	Granite	Y	4–11	5	±1	(1)
2	Shenwo	Carbonate		4–8	6	±several	(2)
3	Danjiangkou	Carbonate Metamorphics	Y	2–5	5–9	±1–2	(3)
4	Dahua	Carbonate	Y	1–7	3.1–3.5	±2	(4)
5	Shengjiaxia	Granite					(5)
6	Shuikou	Granite					(5)
7	Zhelin	Carbonate	Y	3–6	6–7	±several	(6)
8	Qianjin	Carbonate		<1	2	±1	(7)
9	Tongjiezi	Carbonate		1–4	1.2	±1–3	(8)
10	Hunanzhen	Granitoid		0.3–0.4		±0.5	(9)
11	Wujiangdu	Carbonate		<0.5		±0.5	(5)
12	Nanchong	Carbonate		<1			(10)
13	Huangshi	Carbonate		<2			(11)
14	Yantan	Carbonate		1–7		±2	(4)
15	Geheyan	Carbonate	Y	<1			(5)
16	Lubuge	Carbonate		<3		±2	(12)
17	Nanshui	Carbonate					(13)
18	Dengjiaqiao	Carbonate		<0.5			(14)
19	Dongjiang	Carbonate		3–4		±0.5–0.6	(15)

(1) DING et al. (1987), (2) ZHONG et al. (1981), (3) GAO and YING (1980), (4) GUANG (1995), (5) HU et al. (1996), (6) HUANG and KONG (1984), (7) GAO et al. (1984), (8) GUO (1994), (9) HU et al. (1986), (10) HU and CHEN (1979), (11) KONG (1984), (12) JIANG and WEI (1995), (13) XIAO and PAN (1984), (14) LIU and LI (1981), (15) HU et al. (1995).

considered accurate to about 2 km. The hypocenters of the main shocks and the pursuant seismicity at these two locations (1 to 7 km) are located in Carboniferous and Devonian limestones (GUANG, 1995). The macroscopically determined epicenter of the Zhelin earthquake (No. 7) is located in Cambrian limestone, whereas the main shock (inferred depth of 6 to 7 km) and subsequent seismicity (3 to 6 km) are located in Precambrian limestone and clastic rocks (HUANG and KONG, 1984). A portable seismograph network was deployed at Qianjin Reservoir. The seismicity was very shallow with $(S-P)$ times of 0.1 to 0.2 s and was accompanied by loud earthquake sounds. The main shock (~ 2 km depth) and subsequent seismicity (a few hundred meters deep) are located in Proterozoic and Precambrian limestones, respectively (GAO et al., 1984).

The main shock and subsequent seismicity at Tongjiezi (No. 9) were determined by a dense network of portable stations. Tongjiezi Reservoir was impounded on April 15, 1992 and earthquakes occurred the next day near the dam. Earthquakes in Tongjiezi were shallow (1 to 4 km deep). The main shock was 1.2 km deep and located in Carboniferous limestones (GUO, 1994).

The seismicity at Hunanzhen (No. 10) was located using a dense network of portable seismographs. It occurred at depths of a few hundred meters in granitoid rocks (HU et al., 1986). The seismicity at Wujiangdu (No. 11) was located on a dense portable seismic network. It was very shallow, and events with magnitudes ≈ 0 were associated with loud sounds. The seismicity was located in Permian and Triassic limestones (HU et al., 1996). No seismic stations were deployed at Nanchong (No. 12). However from macroscopic data and loud sounds the hypocenter was considered to be in the top 1 km in middle Devonian limestone (HU and CHEN, 1979).

There were no seismic stations at Huangshi (No. 13) and the depths (<2 km) were estimated from macroscopic data. The seismicity lies in Cambrian and Ordovician limestones. Seismicity at Geheyan (No. 15) was recorded on a local seismic network. The seismicity was shallow ($\sim$few hundred meters deep) and occurred in Permian and Triassic limestones (HU et al., 1996).

Lubuge Reservoir (No. 16) was impounded on November 21, 1988, and by the third day after impoundment, earthquakes with M_L 1.0 began to occur. On the fourth day an earthquake with M_L 2.9 occurred accompanied with rumbling noise. The focal depth was 3.0 km (JIANG and WEI, 1995). The seismicity was located on a local network and the depths are considered accurate to about 1 km. The seismicity was located in upper Carboniferous limestones (JIANG and WEI, 1995).

The seismicity at Dengjiaqiao (No. 18) was very shallow and based on macroscopic data. It was located at depths of a few hundred meters in Cambrian limestones.

The seismicity at Dongjiang (No. 19) was located on a dense seismic network. The depths are considered accurate to ~ 0.5 km. The seismicity was located between depths of 3 to 4 km and was located in lower Carboniferous and Devonian limestones (HU et al., 1995).

No data are available for Shengjiaxia (No. 5), Shuikou (No. 6) and Nanshui (No. 17) reservoirs. The lithology in the epicentral area at these locations is granite for the first two and carbonates for Nanshui.

For the 18 reservoirs where RIS occurred after Xinfengjiang, the magnitudes of the main shocks range between M_s 2.2 and 4.8. There were three earthquakes with magnitudes between 4.5 and 4.8. They occurred at Shenwo (M 4.8), Danjiangkou (M 4.7) and Dahua (M 4.5) (Table 2). Very shallow focal depths (<1 km) accompanied by loud sounds were observed at Qianjin (No. 8), Hunanzhen (No. 10), Wujiangdu (No. 11), Nanchong (No. 12) and Dengjiaqiao (No. 18).

The dam heights and reservoir volumes vary significantly from only 12 m and 4×10^{-4} km^3 for Dengjiaqiao (No. 18) to 165 m high for Wujiangdu (No. 11) and 19 km^3 for Danjiangkou Reservoirs (No. 3). In some reservoirs seismicity began soon after impoundment, whereas at others there was a delay in seismicity of a few years after impoundment. For example, seismicity occurred a day after impoundment started at Tongjiezi Reservoir (No. 9) and three days after the impoundment of Lubuge Reservoir (No. 16). Although the depths are not well constrained they were shallow. The main shock was at a depth of 1.2 km at Tongjiezi and shallower than 3 km at Lubuge. These observations suggest that increased pore pressure due to Skempton's effect is the likely mechanism of RIS. On the other end, the initial seismicity at Danjiangkou (No. 3) and Huangshi (No. 13) was two and three years respectively after impoundment (Table 1). The depth of the main shock at Danjiangkou is not well constrained and lies between 5 and 9 km, whereas the pursuant seismicity was shallower (GAO and CHEN, 1981). At Huangshi the seismicity was shallower than 2 km, yet it occurred three years after impoundment (KONG, 1984). These observations suggest that chemical effects probably played an important role in weakening the carbonate rocks in the karstic environment. Seismicity associated with karst cavity collapse was suggested for Zhelin, Huangshi and Dongjiang Reservoirs (Nos. 7, 13 and 19) (KONG, 1984; HUANG and KONG, 1984; HU et al., 1995).

Four Examples

We present information regarding four examples of RIS in China. These include Xinfengjiang Reservoir; the best studied and the location of the largest earthquake. Next we present information pertaining to Danjiangkou Reservoir where the location of RIS was controlled by lithology. Then we present data relative to RIS near the Shenwo Reservoir, where the temporal pattern displayed interesting association with the large earthquake near Haicheng. The fourth example is of the smallest reservoir known to have been associated with RIS.

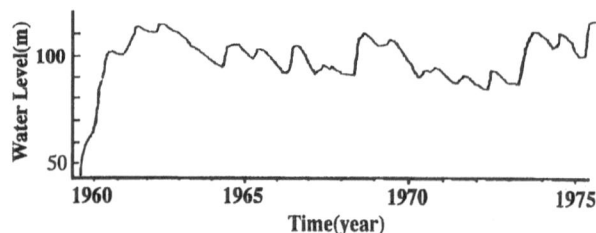

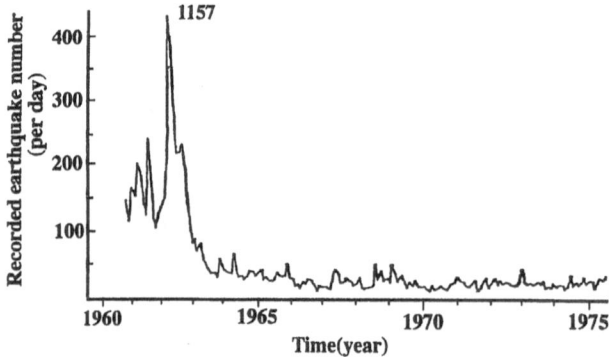

Figure 2
Relation between water level and earthquake frequency at Xinfengjiang Reservoir (modified from DING
et al., 1987). The upper figure shows water level curve and the lower figure shows the earthquake
frequency.

a. Xinfengjiang Reservoir

The RIS at Xinfengjiang Reservoir has been well documented (see for example
WANG *et al.*, 1976; and GUPTA, 1992), and we will present more recent data and
other observations relevant to the mechanism of RIS. The main shock at Xinfengjiang is one of the four examples of RIS worldwide with a magnitude greater than
6.0. The details presented below have been taken from an exhaustive study by DING
et al. (1987). Xinfengjiang dam is one of the four dams in China whose reservoirs
are located in granitic rock. The 105 m high concrete dam impounds a reservoir
with a volume of 11.5 km^3. Impoundment began on October 20, 1959 and
seismicity started a month later. During the first few years (1961–1966), the
earthquake frequency followed the reservoir levels (Fig. 2). The main shock M_s 6.1,
occurred on March 19, 1962.

The dam and reservoir are located in a Mesozoic granitic batholith, with three
well developed sets of faults striking NNW, NNE and NEE (Fig. 3). The main
shock occurred near the intersection area of the NNW and NEE trending faults
(Fig. 3). Joints are well developed in the epicentral area and divide the granite
batholith into several blocks. By the end of 1987, 337,461 earthquakes were

recorded by the reservoir network, 13,643 earthquakes with magnitude $M_s \geq 1.0$, 313 with magnitude $M_s \geq 3.0$ and 49 earthquakes with magnitude $M_s \geq 4.0$. There were two aftershocks with magnitude greater than M_s 5. The main shock registered a magnitude M_s 6.1, the focal depth of 5 km. The main shock encompassed a meizoseismal area of 28 km^2 and epicentral intensity VIII. The main shock caused extensive damage to houses and generated cracks in the river bank. It also generated cracks in the dam.

All the epicenters were near the reservoirs, outside the deepest part of the reservoir within 5 km of the reservoir. The earthquakes were distributed in four areas, area A, B, C and D (Fig. 4). Area A is located in a gorge area downstream of the dam. It is about 12 km long and 8 km wide. More than 90% of the earthquakes occurred in this area, most of the earthquakes with $M_s \geq 3.0$ were also located here. The main shock and the two earthquakes with $M_s > 5.0$ also occurred here. Seismicity continued in this area when the earthquakes stopped in B, C and D areas. Figure 4 displays all $M_s \geq 2.0$ earthquakes that occurred between July,

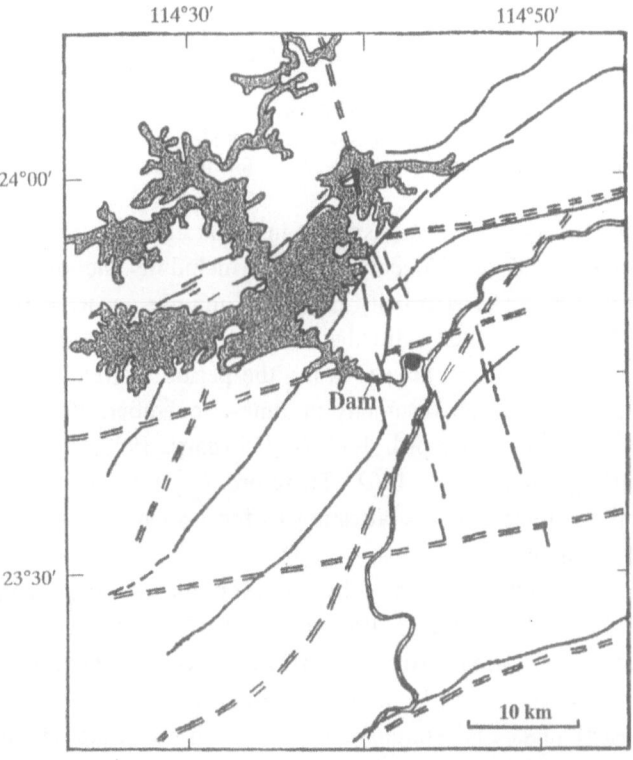

Figure 3

Major faults in the Xinfengjiang Reservoir area. Faults exposed on the surface are shown by single lines, whereas deeper, buried faults are shown by double lines (modified from DING et al., 1987).

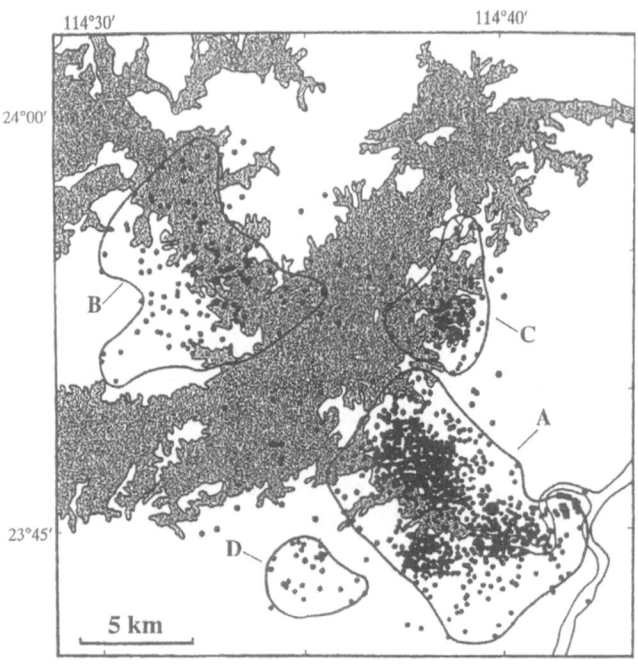

Figure 4
Distribution of earthquakes with magnitude $M_s \geq 2.0$ in Xinfengjiang Reservoir (July 1961–December 1978). The main shock occurred in area A (modified from DING *et al.*, 1987).

1961 and December, 1978. The seismicity in the A area occurred mainly during the period October, 1959 to November, 1962, including the most intense period, March–April, 1962. The seismicity in area B mainly occurred during the period October, 1961 to May, 1962, with the most intense activity in November–December, 1961. The seismicity in area C is for the period October, 1961 to December, 1963 with the most intense seismicity in June–November, 1962. The seismicity in area D covers the period April, 1962 to February, 1963, with the most intense seismicity arising in August, 1962. Therefore we note that the different areas "lighted" up at different times, attesting to the presence of a non-uniform set of fractures in the area.

The epicentral area increased with time. Figure 5 shows the temporal change in the epicentral area. From July 8, 1961 to March 18, 1962, most of the earthquakes were located in the A area. From March 19 to December 31, 1962, there were also earthquakes in areas C and D, and the epicentral area of A more than doubled.

All the earthquakes in Xinfengjiang were shallower than 15 km, the predominant depth being 4–11 km. The hypocentral depths increased with time (DING *et al.*, 1987). After the main shock, the predominant focal depths were 7–8 km before 1976 and 8–9 km after 1976. The location of seismicity on the periphery of the reservoir, distant from the deepest part of the reservoir, and the epicentral and hypocentral growth suggest that RIS was associated with pore pressure diffusion. Based on the assumption that the diffusion of pore pressure was associated with the increase in epicentral area, the hydraulic diffusivity was estimated (see e.g., TALWANI and ACREE, 1984/85). The hydraulic diffusivity was found to be $1.25 \times 10^4 \; cm^2/s$, a value consistent with other locations of RIS (TALWANI and ACREE, 1984/85).

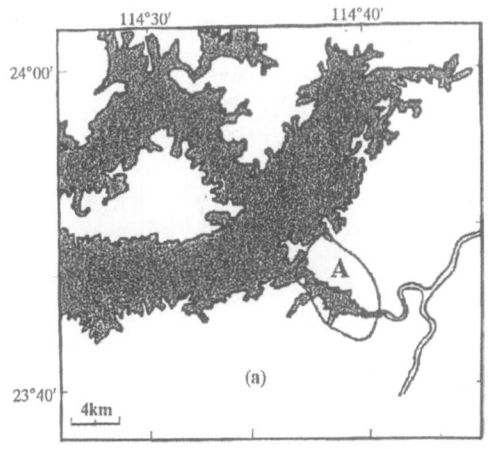

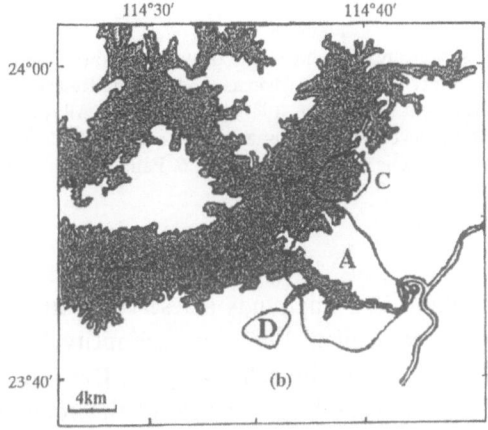

Figure 5
Temporal patterns of seismicity for two time periods in Xinfengjiang Reservoir. (a) July 8, 1961–March 8, 1962; (b) March 19, 1962–December 31, 1962 (modified from DING et al., 1987).

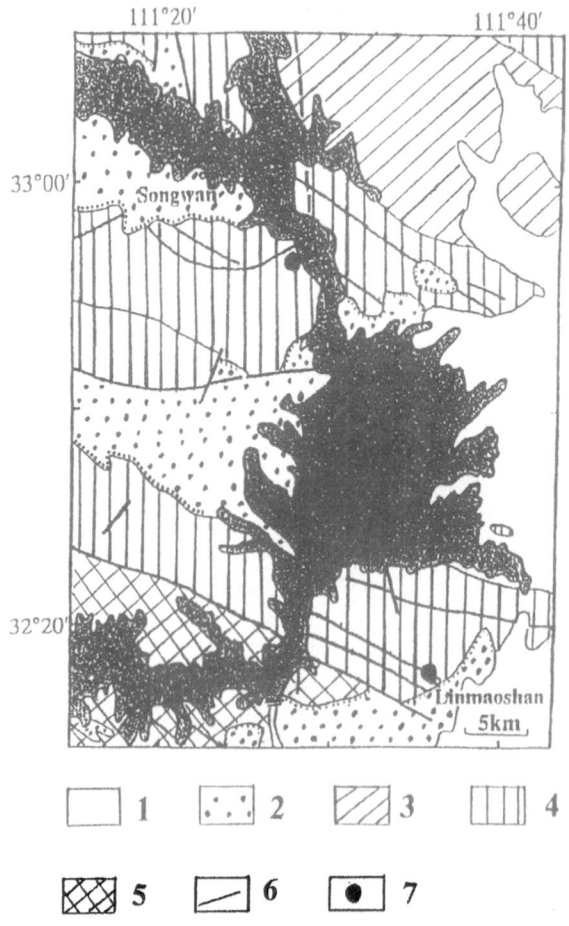

Figure 6

Tectonic map of Danjiangkou Reservoir area which consists of two reservoirs. The Danjiang Reservoir extends from the northwest to the center of the map, whereas the Hanjiang Reservoir extends further to the southwest of the map (modified from DING *et al.*, 1987). 1. Quaternary Alluvium; 2. Eocene-Cretaceous clastic rocks; 3. Carboniferous, Devonian and Silurian clastic rocks; 4. Ordovician, Cambrian and Precambrian carbonate rocks; 5. Proterozoic metamorphic rocks; 6. Fault; 7. Epicenters of main shocks.

b. Danjiangkou Reservoir

The 97-m high Danjiangkou dam impounds a reservoir with a volume of 19 km³. Impoundment began in November 1967, and seismicity began in January 1970. The reservoir is composed of two branches (Fig. 6), Danjiang and Hanjiang Reservoirs to its southwest, occupying 48% and 52% of the volume, respectively. The Hanjiang Reservoir is located on Proterozoic schists and metavolcanic rocks and Cretaceous to Tertiary age sandstones and conglomerates. No earthquakes occurred in the shallower metamorphic terrane in which the Hanjiang Reservoir is located. The main body of the Danjiang Reservoir consists of Cretaceous to Tertiary sandstones, red beds and mudstones. Here also no earthquakes occurred. The epicenters were located in the Ordovician, Cambrian and Precambrian carbonates, especially the karst areas near Songwan Gorge to the northwest and Linmaoshan Gorge to the southeast of the Danjaing Reservoir (Fig. 6). The $M_s = 4.7$, 4.2 and 4.6 earthquakes occurred on November 29, 1973 in the Sonwang area. The depths of these earthquakes were not well constrained at about 9 km, and they could possibly have occurred in the underlying Proterozoic metamorphic rocks. The focal depth of other earthquakes was about 3–5 km or shallower and these occurred in Precambrian dolomites.

This example of RIS at Danjiangkou Reservoir illustrates how the epicentral location of RIS was controlled by the availability of karsts, although the main shock may have occurred in the underlying metamorphic rocks.

c. Shenwo Reservoir

Shenwo Reservoir has a dam height of 50.3 m and impounds a volume of 0.54 km³ (Table 1). Impoundment began in November 1972 and earthquakes initiated in February 1973. The main shock had a magnitude M_s 4.8, and it occurred over 10 km upstream of the dam on December 22, 1974. The focal depths ranged from 4 to 8 km and the main shock reached a focal depth of 6 km. The fault-plane solution showed that faulting occurred along a NE fault in the reservoir.

The epicenter of the Shenwo main shock is about 100 km from the epicenter of the February 4, 1975 Haicheng M 7.3 earthquake (Fig. 7), and both sites are under the same regional stress field. Figure 8 shows the water levels and seismicity in Shenwo Reservoir. We note that there were no earthquakes with magnitude $M \geq 1.0$ in the reservoir area for one month following the Haicheng earthquake.

The Shenwo Reservoir is located in the dilatational quadrant of the Haicheng earthquake. It appears that there was a static stress drop in the Shenwo Reservoir area due to the Haicheng earthquake. KING et al. (1994) demonstrated that the 1992 M 7.3 Landers earthquake caused static stress change of 0.2–0.3 bars in the areas 100 km away. They showed that stress increases of less than one-half bar appeared sufficient to trigger earthquakes, and stress decreases of a similar amount were sufficient to suppress them. Therefore, the small reduced static stress in the Shenwo area, caused by the Haicheng earthquake, may be responsible for the quiescence in seismicity in Shenwo Reservoir area immediately following the Haicheng earthquake. Since the Shenwo main shock occurred about forty days before the Haicheng earthquake, it could also have had a triggering effect on the Haicheng earthquake, although the effect is less because of the smaller magnitude of the Shenwo earthquake.

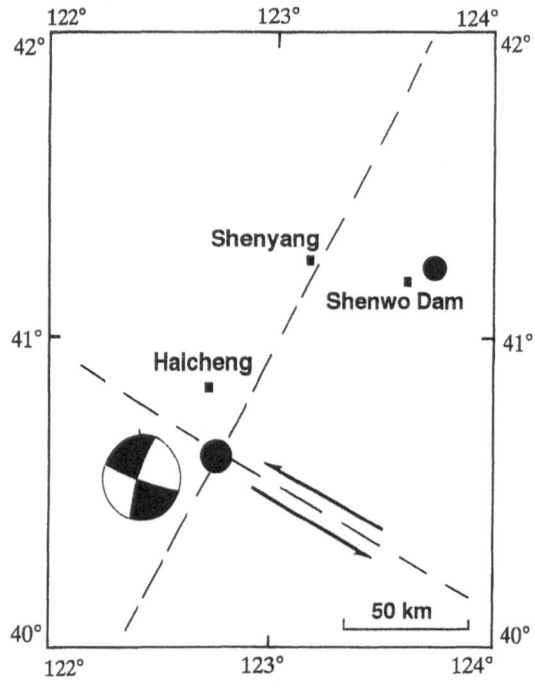

Figure 7
Map showing the location of Haicheng and Shenwo and the strike directions of the nodal planes of
Haicheng earthquake (dashed lines). The fault plane solution is from Wu *et al.* (1979).

d. *Dengjiaqiao Reservoir*

Dengjiaqaio Reservoir has a dam height of 12 m and a volume of 3.5×10^5 m^3
(Table 1). Impoundment began in December 1979 and heavy rain fell in June and

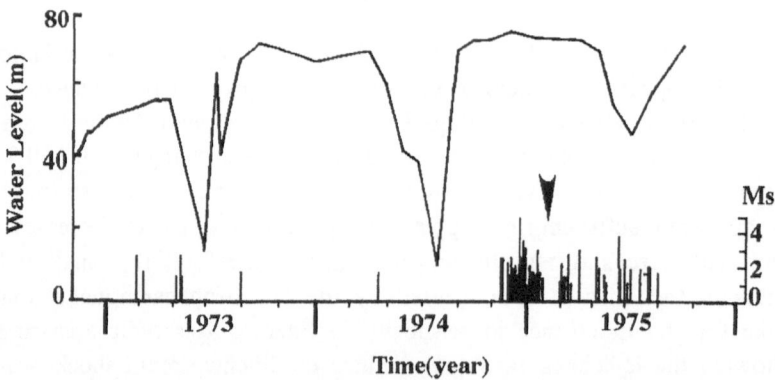

Figure 8
Water level and seismicity in Shenwo Reservoir (modified from ZHONG *et al.*, 1981). The time of
Haicheng earthquake is shown by the thick arrow.

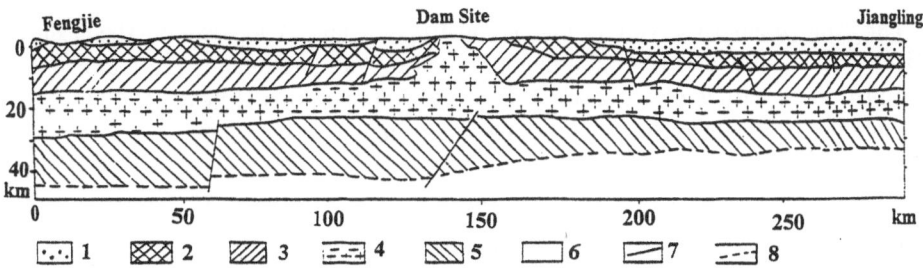

Figure 9

Cross-section from Fenjie to Jiangling along Yangtze River (modified from Yangtze River Water Resource Committee). The Three Gorges dam will be located on the granite core. 1. Alluvium, sandstone and shale; 2. Limestone; 3. Base metamorphic rocks; 4. Granite; 5. Diorite and Gabrro; 6. Olivine; 7. Fault; 8. Mohorovicic boundary (no vertical exaggeration).

July 1980, and the reservoir was full on July 31, 1980. Earthquakes were first felt on August 1, 1980, and a M_L 1.9 earthquake occurred that day. On October 30, 1983, another earthquake with magnitude M_L 2.2 occurred after another sizable rain storm. Both the earthquakes had shallow focal depths and small felt areas, and were associated with loud sounds.

The reservoir is located on the hanging wall of a NE-striking normal fault which is less than 1 km away from the reservoir. The reservoir area consists of limestones, dolomites and dolomitic limestones, with well developed karsts. Both the felt earthquakes occurred shortly after extensive rainfalls. One possible mechanism responsible for the earthquakes is that the loading effect of the reservoir and the substantial rainfall caused the hanging wall of the normal fault to move downward. Another possible mechanism is that there was a collapse of a cavity in the karst.

Seismicity Potential at the Three Gorges Project and Xiaolangdi Project

The Three Gorges Project under construction on the Yangtze River (Fig. 1) will have a dam height of 175 m and a reservoir volume of 39.3 km³. It will be one of the largest multi-purpose projects in the world. The dam will be built in granitic rock. The granitic rock forms a core within carbonate rocks. The upstream area of the reservoir will lie on both the granitic core and on the upstream limestone beds, with an active fault separating them. A cross section from Fengjie (upstream) to Jiangling (downstream) of the Three Gorges Project is shown in Figure 9. We note that the reservoir lies over both granitic rocks and limestones, which have been associated with RIS. The Three Gorges Project lies in an area of ambient seismicity. Thus the empirical evidence suggests that we can anticipate moderate RIS in the reservoir area, especially upstream of the dam.

The Xiaolangdi Reservoir is under construction and is located on the Yellow River (Fig. 1). It will have a dam height of 154 m and a volume of 12.65 km^3 upon completion. A seismic network has already been in operation (LIN *et al.*, 1995), and no significant seismicity has thus far been detected.

Conclusions

There are 19 cases of RIS in China, including the Xinfengjiang Reservoir which was associated with a M_s 6.1 event in 1962. Most of the cases of RIS occurred in South China and are predominantly in karst terrane. The cases of RIS in granitic rocks, e.g., Xinfengjiang Reservoir appear to be caused by pore pressure diffusion in fractured rocks. That lithology controls the location of seismicity is illustrated by the example of RIS in Danjiangkou Reservoir. The temporal association of RIS with filling showed that in some cases, shallow, small earthquakes are associated with reservoir impoundment (Skempton's effect). Several examples illustrate that the chemical effect of water in dissolution is responsible for RIS.

The presence of faults in the granitic core where the Three Gorges Project is under construction, and the presence of outlying carbonate rocks upstream, suggest the possibility of moderate earthquakes when the reservoir is impounded.

REFERENCES

CHEN, L. Y. (1995), *Inducing Factors in Reservoir-induced Seismicity* (in Chinese), Master's Thesis, Institute of Geology, State Seismological Bureau, Beijing.
DING, Y. Z. *et al.*, *Reservoir-induced Seismicity* (in Chinese) (Seismological Press, Beijing 1987), 187 pp.
FYFE, W. S., PRICE, N. J., and THOMPSON, A. B., *Fluids in the Earth's Crust: Their Significance in Metamorphic, Tectonic and Chemical Transport Processes* (Amsterdam–Oxford–New York 1978) 383 pp.
GAO, X. M., and YING, Z. S., *Seismic. activity at Danjiangkou Reservoir in Hanjiang* (in Chinese). In *Danjiangkou Reservoir-induced Seismicity* (Seismological Press, Beijing 1980) pp. 80–89.
GAO, S. J., and CHEN, Y. S. (1981), *Danjiangkou Reservoir-induced Seismicity at Hanjiang* (in Chinese), Acta Seismological Sinica *3*, 23–31.
GAO, S. J., CHEN, Y. S., and KONG, F. J., *Seismic activity at Qianjin Reservoir* (in Chinese). In *Induced Seismicity in China* (Seismological Press, Beijing 1984) pp. 158–161.
GUANG, Y. H. (1995), *Seismicity-induced by Cascade Reservoirs in Dahau, Yantan Hydroelectric Power Station*, Proceedings of the International Symposium on Reservoir-induced Seismicity, Beijing, 157–163.
GUO, M. K. (1994), *Reservoir-induced Seismicity in Tongjiezi Hydropower* (in Chinese), Sichuan Province, Sichuan Earthquake, 12–23.
GUPTA, H. K., *Reservoir-induced earthquakes*. In *Developments in Geotechnical Engineering* (Elsevier 1992), 364 pp.
HU, Y. L., and CHEN, X. C. (1979), *Discussion on the Reservoir-induced Earthquake in China and Some Problems Related to their Origin* (in Chinese), Seismology and Geology *1*, 45–57.
HU, P., CHEN, X. C., and HU, Y. L. (1995), *Induced Seismicity in Dongjiang Reservoir*, Proceedings of the International Symposium on Reservoir-induced Seismicity, Beijing, 142–150.

HU, Y. L., CHEN, X. C., ZHANG, Z. L., MA, W. T., LIU, Z. Y., and LEI, J. (1986), *Induced Seismicity at Hunanzhen Reservoir, Zhejiang Province*, Seismology and Geology 8, 1–26.

HU, Y. L., LIU, Z. Y., YANG, Q. Y., CHEN, X. C., HU, P., MA, W. T., and LEI, J. (1996), *Induced Seismicity at Wujiangdu Reservoir, China: A Case Induced in Karst Area*, Pure appl. geophys., 409–418.

HUANG, D. S., and KONG, F. J., *Discussion on the Induced Earthquakes in Zhelin Reservoir* (in Chinese). In *Induced Seismicity in China* (Seismological Press, Beijing 1984) pp. 124–128.

JIANG, Q. H., and WEI, Z. S. (1995), *Researches on Seismicity Induced by Lubuge Reservoir in Yunnan Province, China*, Proceedings of the International Symposium on Reservoir-induced Seismicity, Beijing, 197–204.

KING, G. C. P., STEIN, R. S., and LIN, J. (1994), *Static Stress Changes in the Triggering of Earthquakes*, Bull. Seismol. Soc. Am. 84, 935–953.

KONG, F. J. (1984), *Karst Collapsing Reservoir-induced Earthquake in Huangshi Reservoir* (in Chinese). In *Induced Seismicity in China* (Seismological Press, Beijing 1984) pp. 154–157.

LIN, X. S., THOMAS, V., LI, H. S., and YE, S. Q. (1995), *Study on Potential Reservoir-induced Seismicity in Xiaolangdi Reservoir*, Proceedings of the International Symposium on Reservoir-induced Seismicity, Beijing, 270–277.

LIU, Z. S., and LI, Y. J. (1981), *Earthquakes of Dengjiaqiao Reservoir in Yidu County, Hubei Province* (in Chinese), Crust Deformation and Earthquake 1, 91–93.

TALWANI, P., and ACREE, S. (1984/85), *Pore-pressure Diffusion and the Mechanism of Reservoir-induced Seismicity*, Pure appl. geophys. 122, 947–965.

TALWANI, P. (1997), *On the Nature of Reservoir-induced Seismicity*, Pure appl. geophys. 150, 473–492.

WANG, M. Y., YANG, M. Y., HU, Y. L., and CHEN, Y. T. (1976), *Mechanism of the Reservoir Impounding Earthquakes at Hisfengkiang and a Preliminary Endeavour to Discuss their Cause*, Eng. Geol. 10, 331–359.

WU, K. T., YUE, M. S., and WU HU, H. Y. (1979), *Certain Characteristics of the Haicheng Earthquake (M = 7.3) Sequence*, China Geophysics 1, 289–308.

XIAO, A. Y., and PAN, J. X., *Analysis on Seismicity Activity at Nanshui Reservoir* (in Chinese). In *Induced Seismicity in China* (Seismological Press, Beijing 1984) pp. 148–153.

YANG, Z. Y., CHENG, Y. Q., and WANG, H. Z., *The Geology of China* (Clarendon Press, Oxford 1986) 303 pp.

ZHONG, Y. Z., JIANG, Y. Q., and HAN, D. Z. (1981), *Discussion on the Seismicity in Shenwo Reservoir Area, Liaoning Province* (in Chinese), Seismology and Geology 3, 59–68.

(Received January 22, 1998, accepted June 11, 1998)

Pure appl. geophys. 153 (1998) 151–162
0033–4553/98/010151–12 $ 1.50 + 0.20/0

| Pure and Applied Geophysics |

Twenty Years Seismic Monitoring of Induced Seismicity in Northern Albania

Betim Muço[1]

Abstract—The northern part of Albania has been the focus of an intense effort by the Seismological Institute of Albania, for in this area two of the country's biggest reservoirs are located. Three years before the impounding of the Fierza reservoir, a four-station network was installed around it. The possibility of induced seismicity continued after the impoundment of the Fierza reservoir in 1978 and the Komani reservoir, in 1985. The seismicity of the zone and some aspects of induced seismicity including: temporal correlation of seismicity with water level changes, spatial patterns in seismicity, frequency-magnitude relations, fault plane solutions etc., are studied during this period. The presence in this zone of a very important transverse fault, the Shkoder-Peja fault, makes the study of induced seismicity from Fierza and Komani reservoirs even more significant. The studies have shown that the impounding of the Fierza and Komani reservoirs has modified the natural course of microearthquake energy release, increasing the number of swarms in this area.

The fluctuation of the water level in these two reservoirs, in due course, is a potentially important factor in the evaluation of seismicity for Northern Albania and especially in the hazard assessment of this region.

Key words: Induced seismicity, seismic activity, seismic energy release, swarms, hazard assessment.

Introduction

Why the filling of reservoirs causes seismic perturbations, is still poorly understood. Nevertheless, the large and world-wide evidence of reservoir-induced seismicity and the studies about it, have given to seismology an interesting topic which continues deservedly to draw the attention of many seismologists.

In Northern Albania there are two large reservoirs which have been seismically monitored since their filling and even before; these are the Fierza and Komani reservoirs, both situated on the Drini river. The upstream end of the Komani reservoir reaches the Fierza dam (see Figs. 1 and 2) so one can consider them as a single, large reservoir, whose total water volume exceeds $4.0 \times 10^9 \, \text{m}^3$.

Three years before the impounding of the Fierza reservoir, a four-station network with short-period instruments was installed around it. These stations are

---

[1] Seismological Institute, Tirana, Albania.

part of the Albanian Seismological Network (ASN). For induced seismicity monitoring in and around Fierza and Komani reservoirs, the data collected from all the stations of ASN, combined with those from other networks such as Montenegro and Macedonia, have been used (Fig. 1). Some studies of the characteristics of the seismic activity of the surrounding areas of both reservoirs have already been carried out and published (MUÇO, 1982, 1985, 1990, 1991a,b, 1992). Figure 2 indicates where these reservoirs are located and the epicenters of earthquakes with $M_L \geq 3.0$ for the period 1976–1995 (MUÇO, 1996).

The Fierza Reservoir

The Fierza reservoir is Albania's largest reservoir and it is also one of the world's largest high dams reservoirs (reservoir volume at maximum water level, 2.8×10^9 m³; dam height, 167 m and dam capacity, 8×10^6 m³).

Though the geological evidence suggests that the studied area has been tectonically active in the recent geological past, this zone has exhibited remarkably low seismicity in historic time (SULSTAROVA et al., 1980). For monitoring earthquake activity before and after the impounding of the Fierza reservoir, the Albanian Seismological Institute has maintained a seismic network there since January 1976. Initially, there were four, three-component short-period seismographs encompassing the reservoir. Afterwards, all the seismological stations around the reservoir were employed in the data analysis (see Fig. 1).

The filling of the reservoir began in October 1978 and the maximum depth of 130 m was attained in April, 1981. Depending on the weather, from 1978 to date, very significant fluctuations of the water level occurred in this reservoir. In the zone surrounding the reservoir there were 138 microearthquakes located before the beginning of the impoundment. There were 305 others for the three first years after the impoundment, from October 1978 to December 1981 (MUÇO, 1982, 1985, 1991a). From January 1982 to December 1992, 1311 earthquakes were recorded and 740 of them were located in this zone (MUÇO, 1992). Comparing the numbers of earthquakes recorded and localized before and after the impoundment of the Komani reservoir, the latter is by far bigger (Fig. 3).

The algorithm used for hypocenter localization is FASTHYPO (HERRMAN, 1979), with a modified procedure for a better depth determination (MUÇO, 1992). In this procedure, using a fixed depth ($H = 10$ km), three other parameters: geographical latitude and longitude (F, L) and origin time (T) are initially determined. Thereafter, with F, L and T fixed, some iterations are carried out to obtain the best solution (the minimum of $r_p = \Sigma(t_i - T_i)^2$, where t_i, T_i are respectively the observed and calculated times for i-th station), changing the depth (H) with a step 5 km and after, 1 km. Considering this solution as a preliminary one, the location procedure recalculates simultaneously all parameters, tending to a minimum of r_p. A single layered model, a constant crust thickness of 40 km and a linear increasing step of velocity: $V = 5.3 + 0.03\ H$, are assumed in the above localization procedure (MUÇO, 1992).

The magnitude determination is made by the formula (MUÇO and MINGA, 1991):

$$M_L = \log(A/T) + 1.804 \log D + 0.0009D - 3.58$$

where A—amplitude in nm, T—period in sec and D—distance-station epicenter in km. Sometimes, magnitude is determined using a signal duration formula (MUÇO and MINGA, 1991):

$$M_D = 2.5 \log \tau - 2.28$$

where τ—signal duration in sec. The earthquake magnitudes in the period January 1976 to December 1981 ranged from 1.3–3.6, while from 1982 to 1992 they reach $M_L = 4.7$. Most earthquake activity has occurred in the form of swarms. Figure 4 shows the cumulative number of earthquakes with $M_L \geq 2.0$ which occurred in and

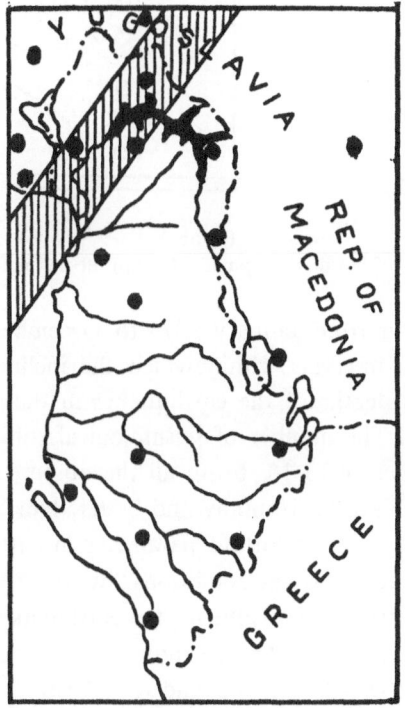

Figure 1
The seismological stations in the region of the reservoirs of Fierza and Komani. The linear shaded area defines the zone of Shkoder-Peja transcurrent fault.

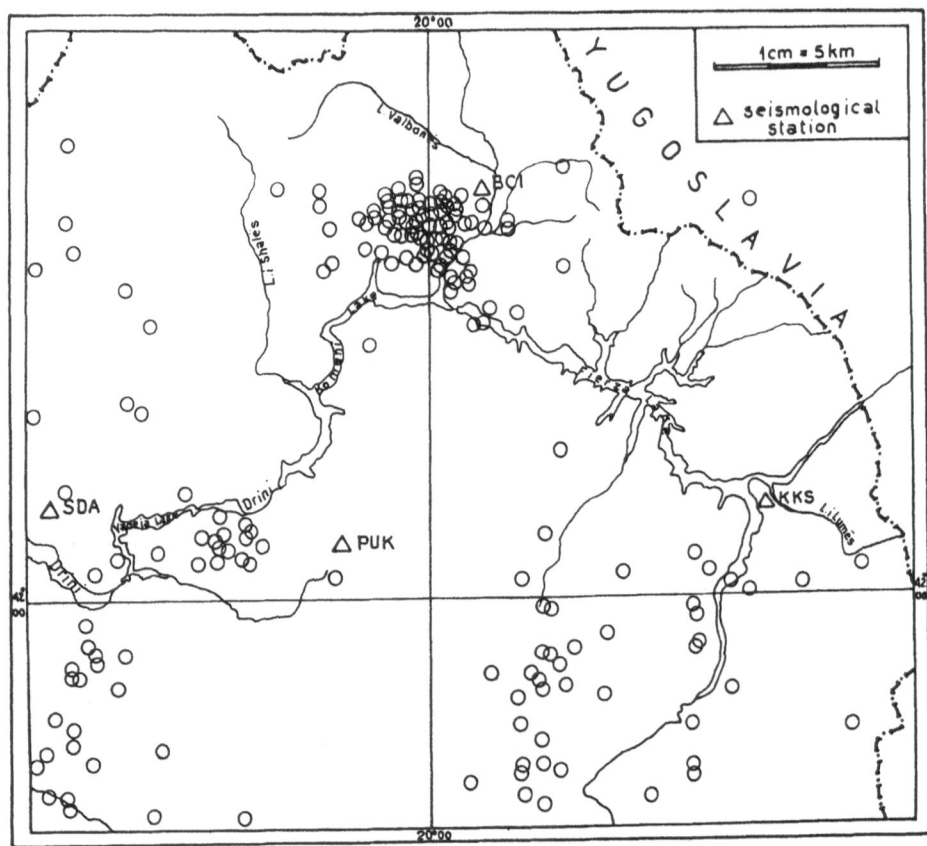

Figure 2
The seismicity of Northern Albania ($M \geq 3.0$) for the period 1976–1995.

near the Fierza reservoir from January 1976 to December 1995. Vertical dashed lines separate the time intervals within which the inclination of graph changes significantly. The focal depths of the earthquakes in the vicinity of the reservoir ranged from 0–12 km. The number of seismological stations used in hypocenter determination varied from 4 to 16. From all the observed events, 93 percent are located to better than 5 km horizontally and better than 7 km for depth.

Because the b value is an important parameter in earthquake hazard analysis and is related to the present stress conditions, by using the least-square method (PEÇI and MINXHOZI, 1988), the b value for the earthquakes of the Fierza reservoir has been monitored continually. The earthquakes are grouped around the mean magnitudes 1.7, 2.2, 2.7, 3.2, 3.7, 4.2 and 4.7. Using a moving one-year time window, shifting it for one month in the time scale; the variation of b value for 193 points is obtained (Fig. 5). The b values before and after impounding are respectively 0.65 ± 0.02 and 0.95 ± 0.01. To verify that the b value change in this case is not accidental, a Fisher test was carried out (WADSWORTH and BRYAN, 1974) producing a positive result at 95 percent conficence level. This is another confirmation that the *b* value of reservoir-associated earthquakes is relatively higher than normally found for the normal earthquakes (GUPTA and RASTOGI, 1976). For the entire observation period, January 1976–December 1992, the frequency-magnitude relation:

$$N = 5.19 + 0.89 M_L$$

is obtained.

Extending our study to the focal mechanism solutions, using the composite focal mechanism, we found the same solutions for 9 groups of earthquakes of Northern Albania for both pre- and post-impoundment periods. For the Bytyçi zone, to the north of the Fierza reservoir, a change of stress tensor is revealed: for the same pair of nodal planes, the tensional and compressional axes before reservoir impoundment were altered after it (Fig. 6) (MUÇO, 1990, 1991a). This is also apparent in other cases (SOBOLEVA, 1980). It is our opinion that alteration of stresses occurs due to the influence of reservoir height and can be observed only on a small local scale. For other parts of Northern Albania there is no evidence of such alterations in tectonic stresses before and after the Fierza reservoir impoundment.

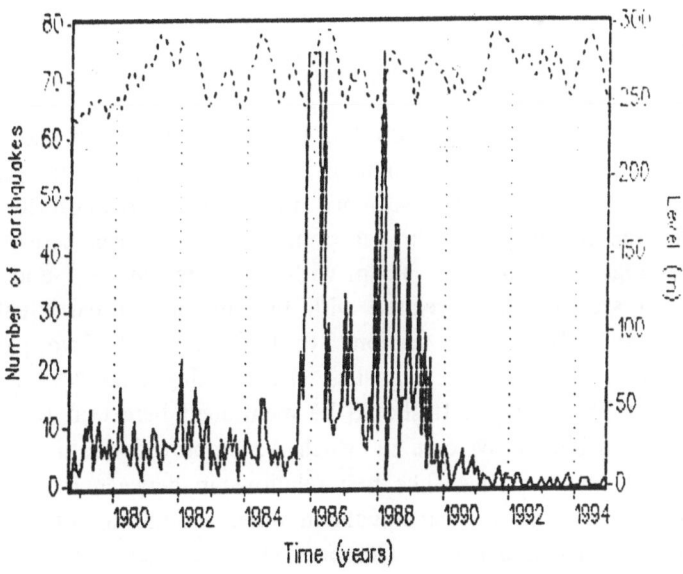

Figure 3
Histogram of monthly numbers of earthquakes ($M \geq 2.0$) near Lake Fierza (solid line) and the variation of its water level, 1976–1994 (dotted line).

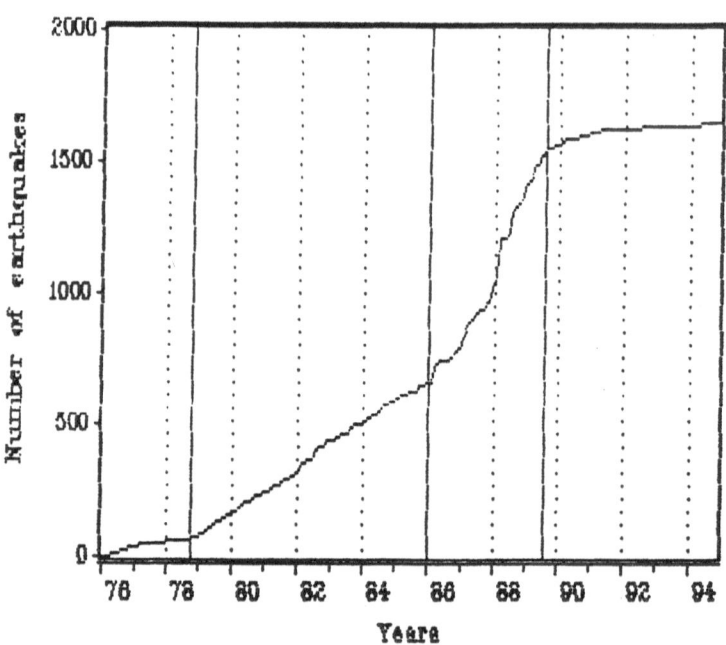

Figure 4
Cumulative number of earthquakes near Lake Fierza, 1976–1994. Solid vertical lines separate time
intervals associated with different seismicity rates.

Previous work (GUPTA and RASTOGI, 1976; SIMPSON, 1976) has suggested a positive correlation between the height of the water column in the reservoir and reservoir-induced seismicity. We searched this cross correlation for the entire period after the impoundment, transforming the water-level series in another one, substituting each change of $+3$ m with 1, -3 m with -1 and so on. The obtained data set is cross-correlated with the number of earthquakes in due course, using the Micro TSP, version 7.0 (LILIEN, 1990). Two pairs of time series were employed for these statistics: the monthly data (195 months) and ten-days data (595 ten days). It should be mentioned here that we achieved no favorable relationship in the case in which we employed the time series in all its length, from 1978 to 1994. The best relationship between the water level in the Fierza reservoir and the earthquake number around it could be obtained only for the period October 1978–October 1985 (the second interval in Fig. 4). The time lags for the best relationship are 1, 2 and 3 months (for the monthly series) and 5 to 7 ten-days (for the ten-days series). These relationships are significant at the 95 percent significance level. The results are shown in Table 1.

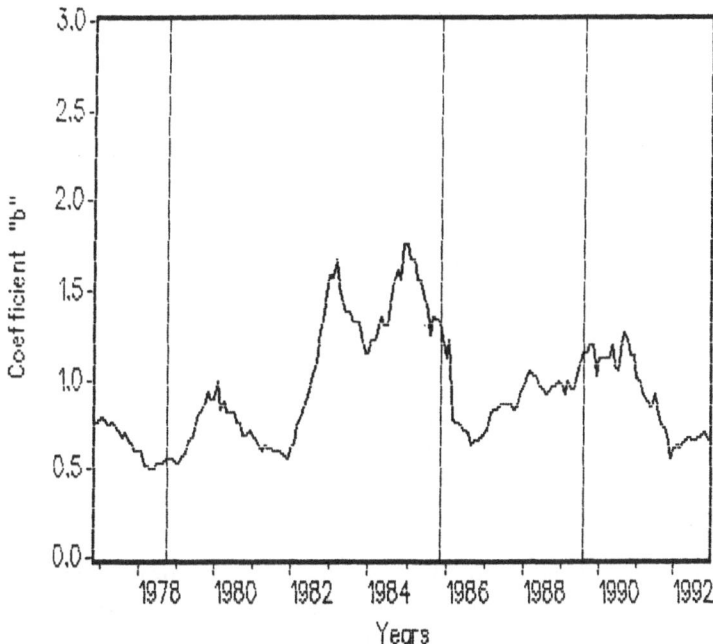

Figure 5
The *b*-value temporal variation for the earthquakes near Lake Fierza, 1976–1992. The vertical lines denote the same time intervals as in Figure 4.

The Komani Reservoir

The Komani reservoir (reservoir volume at maximum water level, 1.6×10^9 m³; dam height 130 m) began to be filled by the end of October 1985. Because of the role of the Fierza reservoir as regulator, the water level of the Komani reservoir has shown only minor changes. Just after the raising of the new reservoir, which

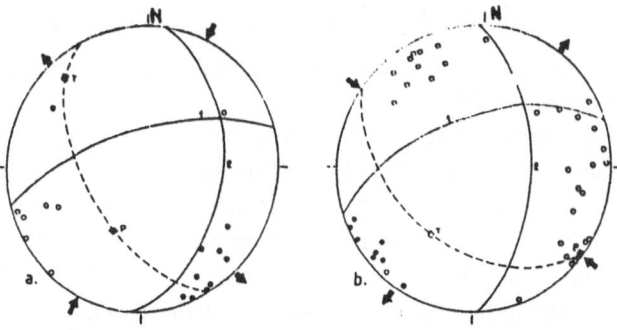

Figure 6
Focal mechanism solutions for earthquakes in the vicinity of the Fierza reservoir, (a) before and (b) after its impoundment, showing alteration of the principal stresses.

Table 1

Cross correlation between monthly series		Cross correlation between ten-days series	
Time lag	Correlation coeff.	Time lag	Correlation coeff.
0	0.174	0	0.034
1	0.245*	1	0.128
2	0.265*	2	0.129
3	0.250*	3	0.149
4	0.193	4	0.153
5	0.112	5	0.161*
6	0.084	6	0.165*
7	0.107	7	0.168*
8	0.062	8	0.146
9	0.028	9	0.146

* The best cross-correlation coefficients.

coincided with a heavy rainfall period, a big microearthquake swarm occurred here. This is the Nikaj-Merturi swarm, which has already been the object of detailed studies (MUÇO, 1989, 1991b, 1992) (Figs. 7, 8). The swarm began on November 10, 1985, grew steadily and the first earthquake with $M_L = 4.0$ occurred on November 19, 1985. The most intensive time was during November–December, 1985 and January, 1986. The swarm drew to its end in August 1986, producing 17096 shocks (recorded at Bajram Curri seismological station, the closest to the epicentral area). The maximum magnitude was $M_L = 4.2$. There were about 300 felt earthquakes and some caused structural damages but no injuries or casualties among the population. The total value of the seismic energy released by the Nikag-Merturi swarm is: $E = 3.94 \times 10^{18}$ erg, which is equivalent to an earthquake of $M_L = 4.9$. The energy is calculated using the formula (RICHTER, 1958):

$$\log E = 9.9 + 1.9 M_L - 0.024 M_L^2.$$

This energy is 52 percent of the entire seismic energy released in Northern Albania from 1976 to 1986.

The area of which the reservoir of Komani is part is inside the well-known transverse fault system, that of Shkoder-Peja (see Fig. 1). Though this fault is very significant from the geological point of view, the evidence for abundant and high-level seismicity in its area is poor. Nevertheless, the expected seismic potential of this zone, which is $M_{max} = 5.5-5.9$ (ALIAJ, 1988), should be taken into consideration in the induced seismicity studies of the Komani reservoir.

The minor changes of the water level in the Komani reservoir did not allow the possibility of examining cross-correlation statistics between water levels and earthquakes.

We also determined focal mechanism solutions for both composite focal mechanism and individual events for the swarm of Nikaj-Merturi. The corresponding tensional and compressional stresses for this entire swarm are shown in Figure 9. The predominant direction of the nearly horizontal tensional axes is NW–SE. The error of the stress direction obtained is $\pm 10°$. The solutions show that fault mechanisms are of normal type (with a distinct strike-slip component) and the stress field corresponds to the regional one (Muço, 1992b).

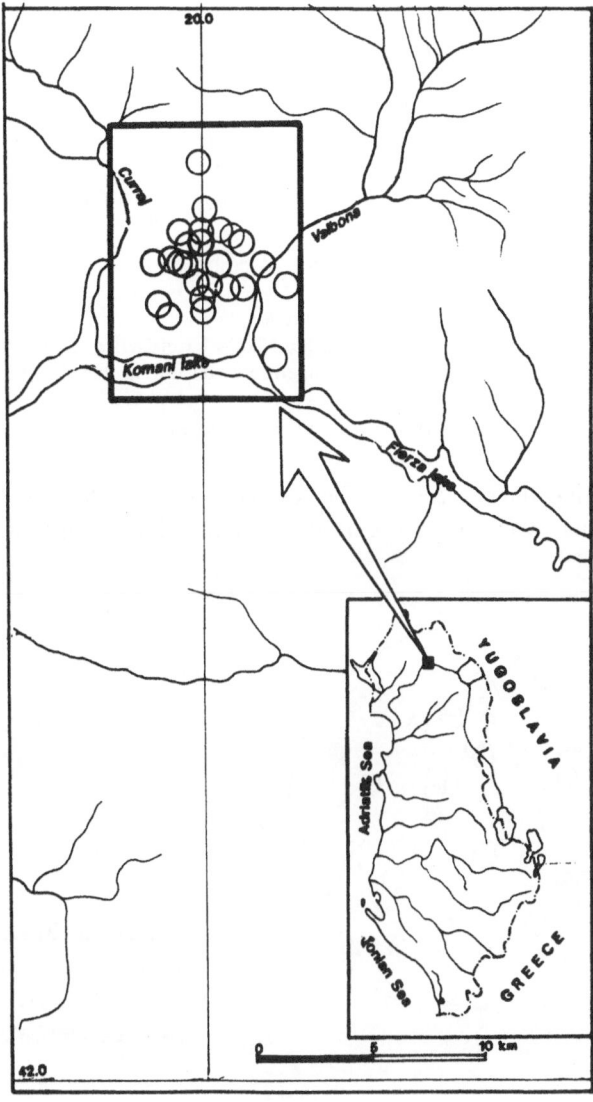

Figure 7
Location of earthquakes in the Nikaj-Merturi series ($M \geq 3.0$).

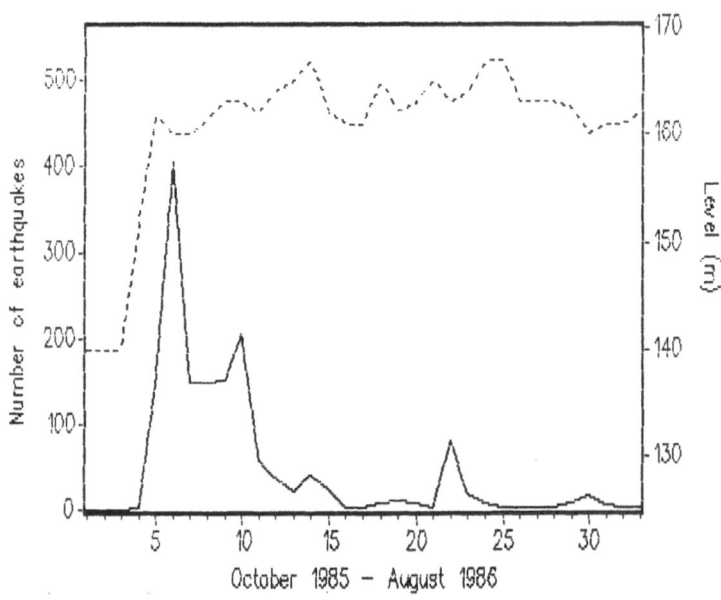

Figure 8
The histogram of ten-day sums of earthquakes in the Nikaj-Merturi series, October 1985–August 1986
(solid line) and the water level of the Komani reservoir during that time (dotted line).

Discussion and Conclusions

We believe that the presence of the reservoir has only hastened the arrival of
seismic events that would have occurred at a latter time regardless. In zones that

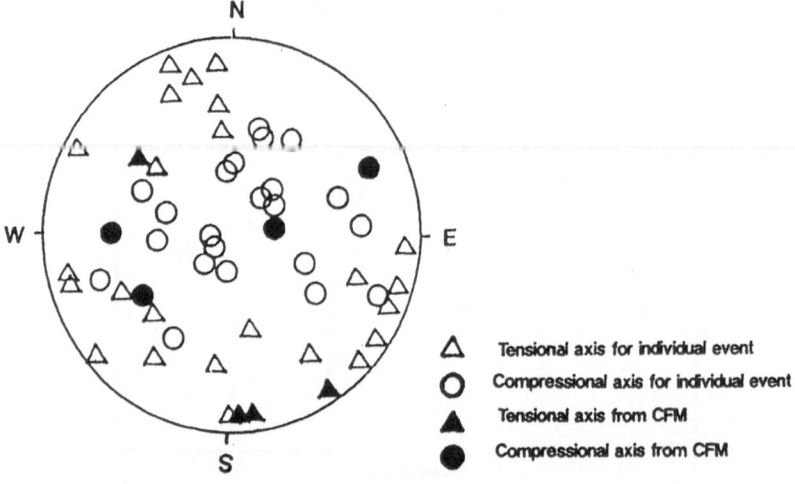

Figure 9
Distribution of stress axes obtanied from the focal mechanism solutions of earthquakes of the
Nikaj-Merturi series, November 1985–August 1986.

have exhibited low seismicity, the influence of reservoir impoundment on their seismicity may be more distinct than in the other ones with a high rate of stress accumulation (SIMPSON, 1976). This is the case in Northern Albania.

Twenty years of seismic monitoring in Northern Albania have revealed interesting seismic phenomena which we think are related to the filling of two reservoirs: Fierza and Komani. The observations have shown that the number of microearthquakes have increased after the impoundment, and many swarms have occurred in the vicinity of these reservoirs.

We believe that the impounding of the Fierza and Komani reservoirs has modified the natural course of microearthquakes energy release. In the case of the Fierza reservoir we could say that the induced seismicity is apparent mostly northward of the reservoir. Analyzing the cumulative number of earthquakes with epicenters in and around the Fierza reservoir (Fig. 4), one can distinguish four intervals. The first one begins in 1976, enduring to the end of 1978. The second continues to the end of 1985. There is a clear difference in the inclination between these two segments which define a higher rate of seismic activity during the second interval when the Fierza reservoir has already been impounded. A clear relationship between the water level and the number of earthquakes in and around the reservoir is obtained for this interval. The time lag for the best cross correlation is from one to three months, which corresponds to the water percolation and time for transmitting the increase of pore pressure to the source depth. The third interval is from the end of 1985 to mid-1989. The rate of seismic activity during this time increased even more and we believe that it is due to the influence of the Komani reservoir, whose impoundment began in October 1985. There is no clear cross correlation between water level and earthquakes for this time interval because the influence of the combined Komani and Fierza reservoirs is very complex. The fourth and last interval (mid-1989 to 1995) has a very similar inclination to the first one, the pre-impoundment period as one can see from Figure 4. Is it a sign that the rocks under the reservoir have stabilized with both loading and fluid pore-pressure effects of the reservoir? We are not quite certain.

For the reservoir of Komani, following the Nikaj-Merturi swarm, November 1985–August 1986, to the end of 1992 this epicentral zone was reactivated occasionally in the form of small swarms.

The study of induced seismicity in and around the Fierza and Komani reservoirs continues. The upgrading of existing stations and provision for new ones in this region would provide better evidence of the influence of the above reservoirs on the seismicity, and in particular provide insights into induced seismicity caused by large reservoirs. It could also aid in the hazard assessment of the studied region.

Acknowledgment

The author would like to express his sincere thanks to the entire staff of the Albanian Seismological Network for making possible the seismic observations necessary for this study. He is very grateful to the anonymous reviewer for his valuable comments and to D. Booth for reading the manuscript and assisting in its improvement.

REFERENCES

ALIAJ, SH. (1988), *The Neotectonics and Seismotectonics of Albania*, D.Sc. Thesis, Seismological Institute, Tirana, Albania (in Albanian).

GUPTA, H. K., and RASTOGI, B. K., *Dams and Earthquakes* (Elsevier, Amsterdam 1976) pp. 229.

HERRMAN, R. B. (1979), *Fasthypo—A Hypocenter Location Program*, Earthq. Notes *50*, 64–83.

MUÇO, B. (1982), *The Seismicity of Drini River Valley and the Influence on it of Lake Fierza*, M.Sc. Thesis, Seismological Center, Tirana, Albania, 150 (in Albanian).

MUÇO, B. (1985), *Lake Fierza Lake and its Influence on the Seismicity of its Surrounding Area*, Bul. Shk. Gjeol. *3*, 67–79 (in Albanian with abstract in English).

MUÇO, B. (1989), *The Nikaj–Merturi Series of Earthquakes, November 1985–August 1986*, Report 11, Seismological Center, Tirana, Albania, 100 (in Albanian with an extended abstract in English).

MUÇO, B. (1990), *The Reservoir of Fierza, Northern Albania and its Influence on the Seismicity of the Surrounding Area*, Programme and Abstract. The Seismol. Soc. of Japan *2*, 147.

MUÇO, B. (1991a), *Evidence for Induced Seismicity of Fierza Reservoir, Northern Albania*, Pure appl. geophys. *136*, 265–279.

MUÇO, B. (1991b), *The Swarm of Nikaj-Merturi, Albania*, Bull. Seismol. Soc. Am. *81*, 1015–1021.

MUÇO, B. (1992), *Features of Albanian Earthquakes and the Role of Underground Waters in their Generation*, D.Sc. Thesis. Seismological Center, Tirana, Albania, pp. 210 (in Albanian).

MUÇO, B. (1994), *Focal Mechanism Solutions for Albanian Earthquakes for the Years 1964–1988*, Tectonophysics *23*, 311–323.

PEÇI, V., and MINXHOZI, A. (1988), *On Determination of b value with Cumulative Distribution According to Magnitude*, Bul. Shk. Nat. *1*, 43–52 (in Albanian with abstract in English).

SIMPSON, D. W. (1976), *Seismicity Changes Associated with Reservoir Loading*, Eng. Geol. *10*, 123–150.

SOBOLEVA, O. V. (1980), *Izmenienie mehanizmov ocagov sllabijh zemljetrasenij pod vlijanijem Nurekskovo vodohranilishça*, Fiz. Zemli *1*, 34–42.

SULSTAROVA, E., KOÇIAJ, S., and ALIAJ, SH. (1980), *The Seismic Regionalisation of Albania*, Academy of Sciences, Tirana, pp. 220.

WADSWORTH, G. P., and BRYAN, J. G., *Application of Probability and Random Variables* (John Wiley, New York 1974) pp. 256.

(Received June 24, 1997, revised November 26, 1997, accepted December 12, 1997)

Pure appl. geophys. 153 (1998) 163–177
0033–4553/98/010163–15 $ 1.50 + 0.20/0

Pure and Applied Geophysics

A Frequency-dependent Relation of Coda Q_c for Koyna-Warna Region, India

PRANTIK MANDAL[1] and B. K. RASTOGI[1]

Abstract—Attenuation properties of the lithosphere around the Koyna-Warna seismic zone is studied by estimating the coda-Q_c from 30 local earthquakes of magnitude varying from 1.5 to 3.8. An average lapse time of 65 sec used in the single scattering model sampled a circular area with an average radius of 114 km. The estimated Q_c values show a frequency-dependent relation, $Q_c = 169 \, f^{0.77}$, and range from 169 at 1 Hz to 1565 at 18 Hz. A comparison of worldwide Q studies reveals that for a large frequency range the Q for active regions is low as compared to that for stable regions. However, South Carolina and Norway are exceptions in that their Q is low in the low frequency range while New England and North Iberia are exceptions as they have a Q value similar to that for active regions like Spain, Turkey, Italy and Garhwal Himalaya (STIH), in the higher frequency range. In contrast to this, the Q for the Koyna-Warna area, which belongs to a stable region, is low in the entire frequency range as compared to the stable regions and similar to the active STIH regions.

Key words: Coda-Q, attenuation, Western India.

Introduction

Knowledge of the attenuation property of the medium is an important aspect in studying the earthquake magnitude and fall-off of ground motion with distance which helps in earthquake hazard mitigation. The decay of seismic wave amplitude with distance defines the attenuation of the medium resulting from the conversion of elastic energy to heat (intrinsic attenuation due to heterogeneities which may be caused by sliding along grain boundaries) and the scattering of seismic waves caused by heterogeneities of varied scale inside the earth's interior. The heterogeneities are due to irregular subsurface geometry, velocity perturbations caused by changes in rock type, cracks and faults. The attenuation due to geometrical spreading is a part of the intrinsic attenuation and is a purely elastic process. The inverse of intrinsic attenuation is termed as the intrinsic quality factor, Q_i (due to internal heating), whereas the inverse of scattering attenuation is known as the

---

[1] National Geophysical Research Institute, Hyderabad-500007, India.

scattering quality factor, Q_s (due to scattering) (WU and AKI, 1985). These two quality factors combined give the total quality factor (Q_c) of the medium.

There are several techniques proposed for estimating the Q_c using either a single or multiple scattering model (AKI and CHOUET, 1975; SATO, 1977; ROECKER et al., 1982; FRANKEL and WENNERBERG, 1987; VAN ECK, 1988; JIN and AKI, 1988; ZENG, 1991; HELLWEG et al., 1995). The single scattering model generally addresses the attenuation of the backscattered body waves but not the primary waves, and it mainly provides the estimates of Q_i (PUJADES et al., 1991). While the multiple scattering model provides information regarding both primary and secondary wave attenuations, and it gives the estimates of Q_i as well as Q_s (GAO et al., 1983). A comparison study between the single scattering model and multiple scattering model revealed that the former model is not sufficient to explain the decay of coda amplitude observed in synthetics whereas the latter technique can explain the same (FRANKEL and WENNERBERG, 1987). However, they found the same Q_i for Anza, California using both single and multiple scattering methods. However, GAO et al. (1983) showed that the single scattering method fails for the long lapse time where multiple scattering dominates. Although, SATO (1988) demonstrated, through the modelling of coda-Q_c by considering the concept of fractal distribution for the dispositions of the scatterers, that the single scattering model is valid even for the long lapse time. Recently, SATO (1994) proposed a hybrid technique, combining the analytical solutions corresponding to single scattering and numerical calculations for multiple scattering, for synthesizing the seismogram to obtain better explanations for the effect of attenuation on different seismic phases. However, the single scattering method has been most widely used due to its easier technique. Moreover, FRANKEL and WENNERBERG (1987) have found similar values from single and multiple scattering methods. Hence, we have applied the single scattering model of AKI and CHOUET (1975).

The Koyna-Warna seismic zone of Maharashtra in India, has been seismically active since the initial impoundment of the Koyna Dam in 1962. The M_s 6.3 Koyna earthquake of 1967 destroyed the Koynanagar township and took 200 lives. Damaging earthquakes of $M \geq 5$ continue to occur and one or more earthquakes of $M \geq 4$ strike every year. Up to 1997 over 150 earthquakes of magnitude exceeding 4 and ten earthquakes of magnitude exceeding 5 took place in this area. Thus, a study of the attenuation property of the medium would be of immense importance in assessing the seismic hazard in this seismically active area as to date no information exists on this aspect. In 1994 one broadband and four short-period 3-component digital seismographs were deployed in the region. The recent deployment of digital seismic instruments in the Koyna area now enables us to study the coda-Q_c of the area. The results are compared with the Q_c values for the other regions of the world.

Tectonic Setting

The Koyna-Warna region is situated in the Western Ghats Region of the Deccan volcanic province. The region is characterized by extensive lava flows of the Cretaceous/Tertiary age and flat topped hillocks. These basalt rocks could be around 1–2 km thick. The dominant mode of deformation in the Koyna area is left-lateral strike-slip faulting along a NNE-SSW trending fault (TANDON and CHAUDHURY, 1969; SYKES, 1970) with northward compression as revealed from the fault plane solution for the M 6.3 earthquake of 1967. The composite fault plane solution for the 1995 earthquakes also showed a mechanism similar to that of the 1967 Koyna earthquake (CHADHA et al., 1997). In the southern part of the Koyna seismic zone, normal faulting is observed along a NW-SE trend (RASTOGI and TALWANI, 1980; GUPTA et al., 1980; RASTOGI, 1992).

Data

The seismicity near Warna during September 1993 to December 1995 has shown a prominent concentration after impoundment to 60 m height of the Warna reservoir (RASTOGI et al., 1997). After the start of the recent burst of earthquakes during 1993, a close seismic network consisting of five portable digital recorders (PDAS) (four with 3-component short-period seismometers and one with 3-component broadband seismometers), five 3-component digital accelerographs, and four analog portacorders were deployed around the epicentral area between Koyna and Warna reservoirs in addition to an existing network of eleven analog stations of Maharashtra Engineering Research Institute, Nasik (MERI). Typical epicentral distances to the digital stations and most of the analog stations were 6 to 25 km, increasing to 90 km for a few analog stations. During 1993–96, 9 earthquakes of magnitude $M \geq 4.0$ and two earthquakes of magnitude $M \geq 5.0$ have occurred in the area. A total of about 2000 earthquakes (most of $M \geq 2$ and a good percentage of smaller ones down to M 0.1) were located using HYPO71PC program and phase data from 19 stations operated by the National Geophysical Research Institute, Hyderabad and the Maharashtra Engineering Research Institute, Nasik. The majority of the hypocenters are found to be of B and C quality and the root mean square error of the 'P' residual times are less than 0.2 s. The statistical error estimation of HYPO71PC program reveals that the accuracy in the horizontal dimension is <1 km and that in depth is <2 km for a majority of the earthquakes. Further the accuracy in depth determination has been checked by other methods such as local earthquake moment tensor analysis (manuscript under preparation) and relocation of quarry blasts for bauxite mining, 8 km west of Warna dam (S. S. Rai, personal communication). However, the actual error in depth determination may be slightly more for some events. Figure 1 shows the epicentral map of earthquakes of magnitude ranging from 1 to 4.7 which occurred during 1995.

For Q_c estimation, the data set consists of digital seismograms of 30 earth-quakes of local duration magnitude varying from 1.5 to 3.8 of 1995, whose epicenters are shown in Figure 2. These earthquakes were recorded on one to three digital stations with sufficient coda duration. However, the events were located using first P- and/or S-wave arrivals from over 15 stations (analog as well as digital shown in Fig. 2). For digital stations, a sampling frequency of 100 Hz was used with a 16-bit resolution. At the analog stations, a drum speed of 60 mm/sec was used for the smoked paper recording.

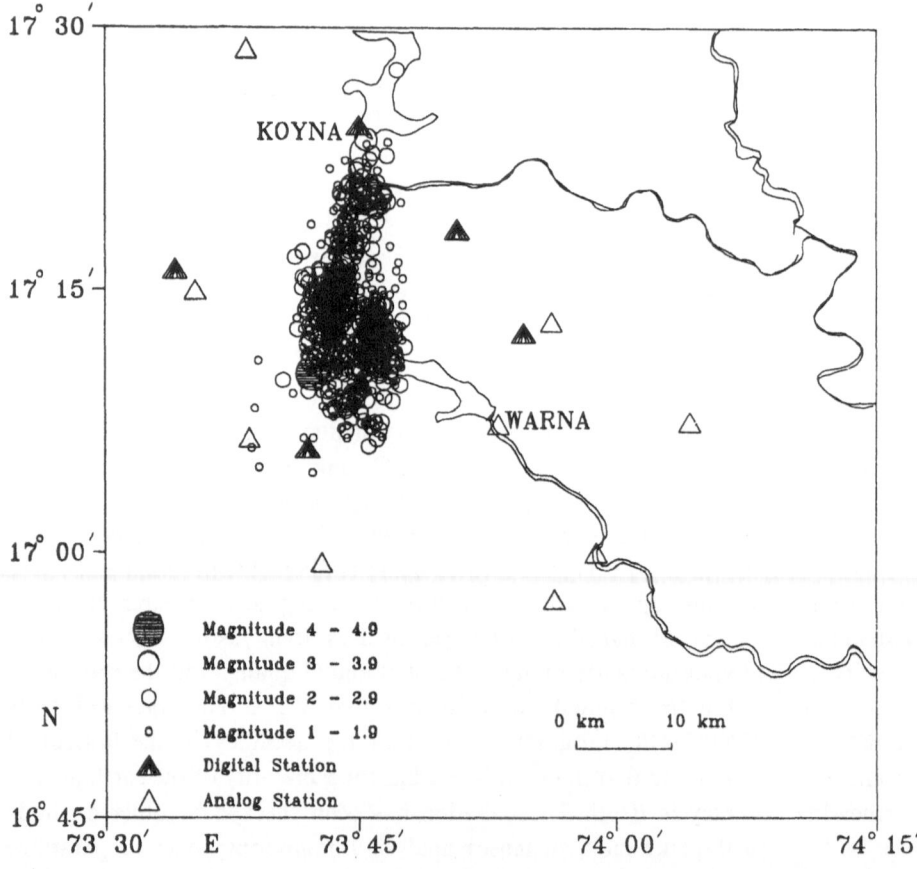

Figure 1
Epicentral map of the Koyna earthquakes which have occurred during 1995. Four analog stations situated at 10–20 km away from the boundary of the figure are not shown. These are well distributed in azimuth.

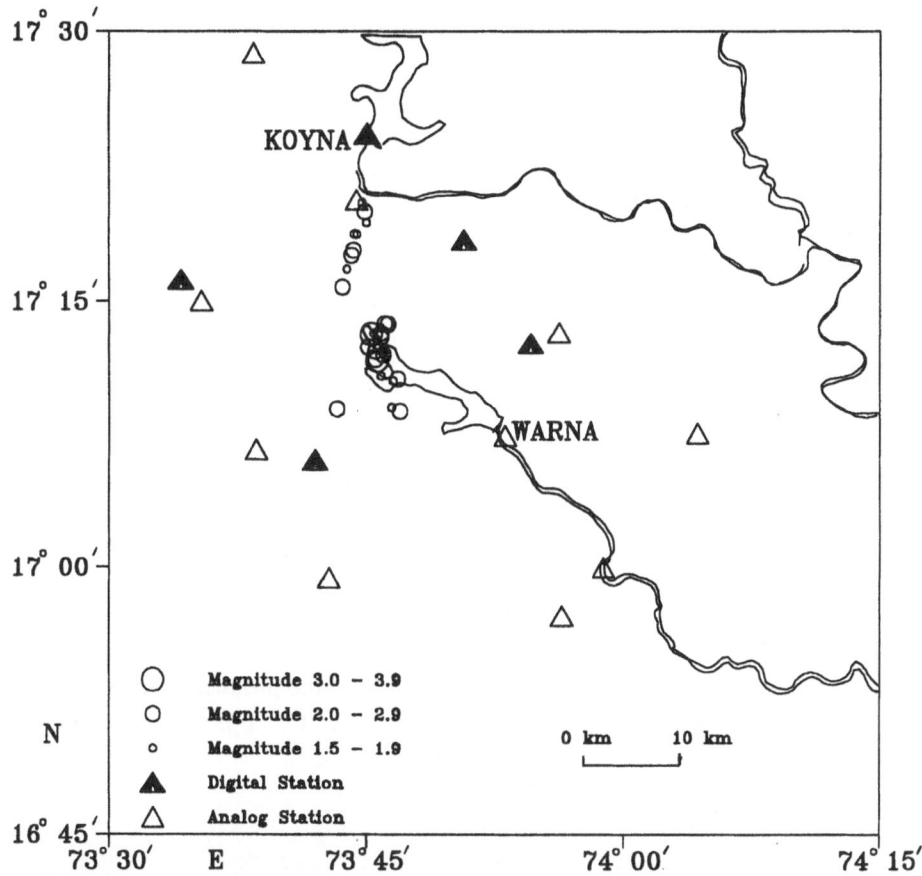

Figure 2
Location map of the selected earthquakes for coda Q_c studies.

Methodology

Coda-Q_c for small local earthquakes is usually estimated with the single scattering method by assuming the coda wave as a combination of backscattered body waves generated by numerous heterogeneities present in the earth's crust and upper mantle (AKI and CHOUET, 1975). In this method, it is further assumed that the source and receiver are colocated. Therefore, implying the scattering as a weak process, the estimated Q_c provides information regarding the intrinsic attenuation property of the medium. Under these assumptions, for a narrow bandwidth signal with a central frequency f_m, the coda-wave amplitude $A(f, t)$ can be expressed as

$$A(f, t) = S(f)t^{-\alpha} \exp\left(\frac{-\pi f t}{Q_c(f)}\right) \tag{1}$$

where t represents the lapse time measured from the origin time of the event. $S(f)$ is the source function, which is assumed to be independent of time and radiation pattern. Therefore, it behaves as a constant. The value a corresponds to the exponent for geometrical spreading and equals 1 for body waves, and Q_c represents the quality factor of the medium.

Logarithm of the equation (1) yields

$$\ln[A(f, t)t] = c - bt \qquad (2)$$

where b and c are equal to $-\pi f/Q_c$ and $\ln(S(f))$, respectively. The slope of equation (2) gives $(1/Q_c)$. Since Q_c is dependent on frequency, the seismograms were filtered using a Butterworth bandpass filter for six different frequency bands (1–2, 2–4, 4–8, 6–12, 8–16, and 12–24 Hz) with central frequencies of 1.5, 3, 6, 9, 12, and 18 Hz, respectively. The beginning of coda is considered at $2t_s$ where t_s is the S-wave travel time from the origin time (RAUTIAN and KHALTURIN, 1978). A variable time window of 50–120 sec, starting from $2t_s$ sec, has been used. Due to variation in magnitude, only 5 of 30 earthquakes have a coda length exceeding 80 sec, while coda length for the remaining earthquakes varies between 50–80 sec. For Q_c estimation, a moving window of 4 sec is used to calculate the RMS average of the filtered coda amplitude which slides in steps of 2 sec. In this process, a smoothed coda envelope is obtained. The logarithm of the product of the smoothed coda envelope and lapse time (t) is drawn. The slope of this plot gives the $(1/Q_c)$ at every central frequency.

Results

The plots between $\ln[A(f, t)t]$ and t for different central frequencies for an earthquake of magnitude M 3.8 are shown in Figure 3. The slope of these plots provides the Q_c values for different frequency bands. As such, these plots suggest that the Q_c value increases with increasing frequency. The estimated Q_c values for different frequency bands are plotted against the different central frequencies (Fig. 4). Next, a power law fitting is used to obtain the frequency dependent coda Q_c relation (i.e., $Q_c = Q_o f^n$, where Q_o and n represent the iso-Q and the slope of the power law fitting) for the region. The estimated Q_c values show a variation of mean Q_c value from 169 at 1 Hz to 1565 at 18 Hz. The frequency dependent coda-Q_c relation for the Koyna region is found to be as given below

$$Q_c = 169 f^{0.77}. \qquad (3)$$

This estimation assumes an isotropic medium. However, in reality, the attenuation property of the medium will vary laterally and vertically due to the anisotropic nature of the lithosphere. Thus, the assumption of the isotropic medium would lead to errors in Q_c and n estimation. The errors in estimated (Q_o, n) values are obtained using the method as described by TOPPING (1963). In this method, we estimate the deviation in Q_o and n using the chi-square technique. The obtained errors $\delta(Q_o)$

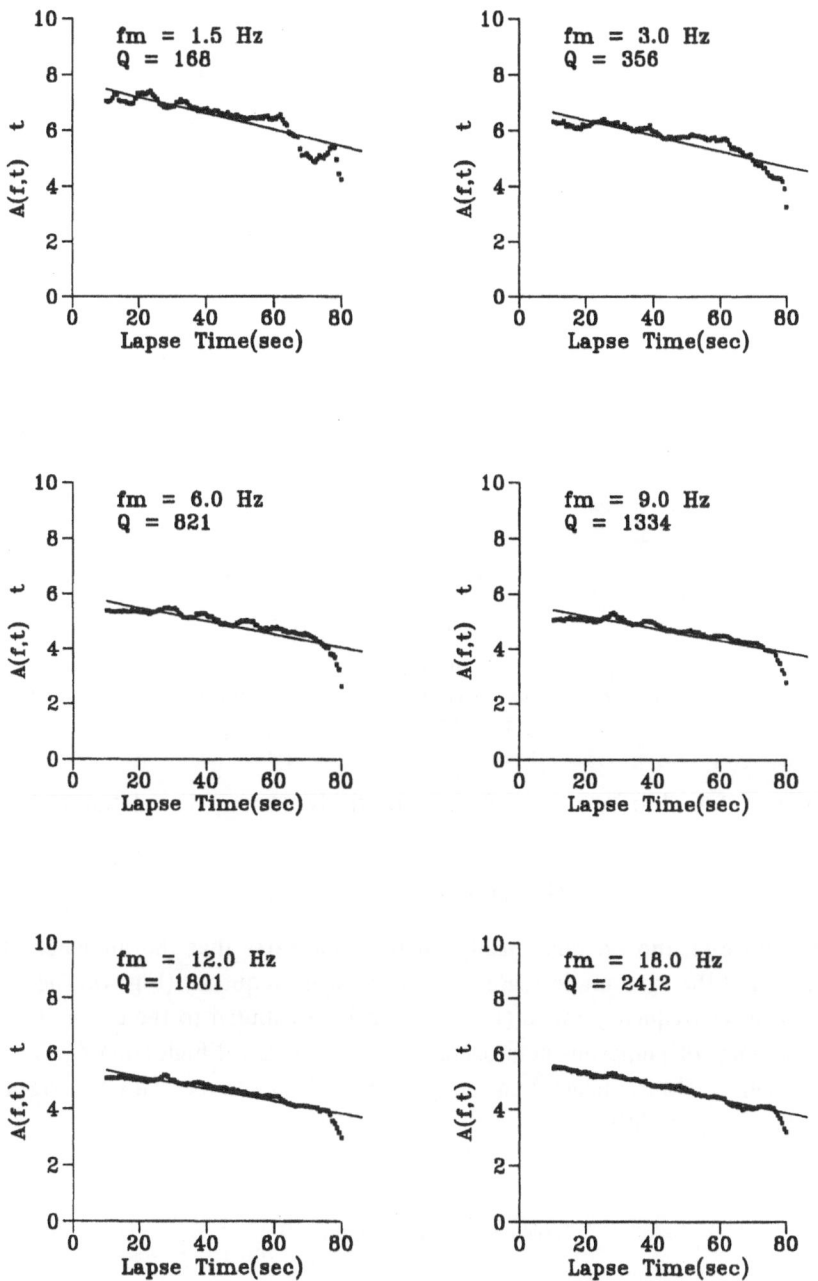

Figure 3
$\ln[A(f, t)*t]$ vs. lapse time (t) plot at different central frequencies for an earthquake of magnitude 3.8.

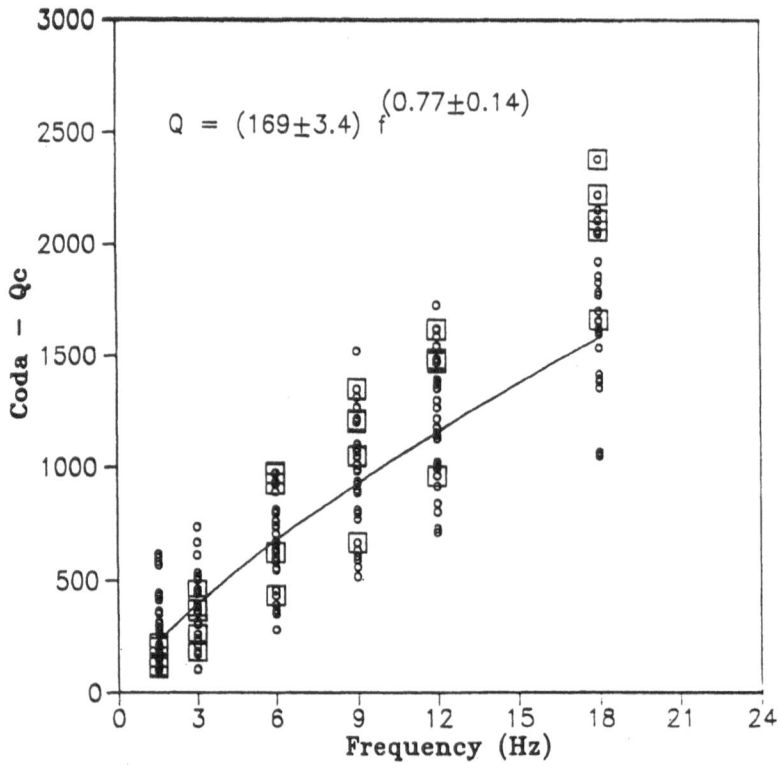

Figure 4

Q_c vs. central frequency (f_m) plot for the Koyna seismic zone. The values with coda lengths of over 80 sec are shown by squares.

and $\delta(n)$ are found to be 3.4 and 0.14, respectively. Hence, the relation (3) should be written as

$$Q_c = (1.69 \pm 3.4) f^{0.77 \pm 0.14}. \tag{4}$$

The above-mentioned relationship, in turn, indicates that the influence of the attenuation of the medium decreases with increase in frequency (Fig. 4). The low Q_c values at lower frequency range (1–3 Hz) can be attributed to the energy loss due to the presence of numerous heterogeneities. The estimated high coda Q_c values at greater than 12 Hz frequency band may be caused by the relatively homogeneous deeper layers in the lithosphere.

Area Covered by Coda-Q_c

The coda-Q_c values represent an average medium property. PULLI (1984) has shown that the scatterers responsible for the generation of coda waves are generally assumed to be distributed over the surface area of an ellipsoid which can be calculated using the following formula:

$$X^2/(vt/2)^2 + Y^2/[vt/2 - R]^2 = 1 \qquad (5)$$

where R equals zero for the coda-Q_c method of AKI and CHOUET (1975). This makes the above equation an equation for the circular area of radius $vt/2$. The parameters v and t represent the velocity of S waves and lapse time, respectively. We have used an average lapse time duration of 65 sec and a S-wave velocity of 3.5 km/sec. Therefore, our coda estimation suggests approximately a circular area of 40,000 km² with a radius of 114 km for the coda wave generation.

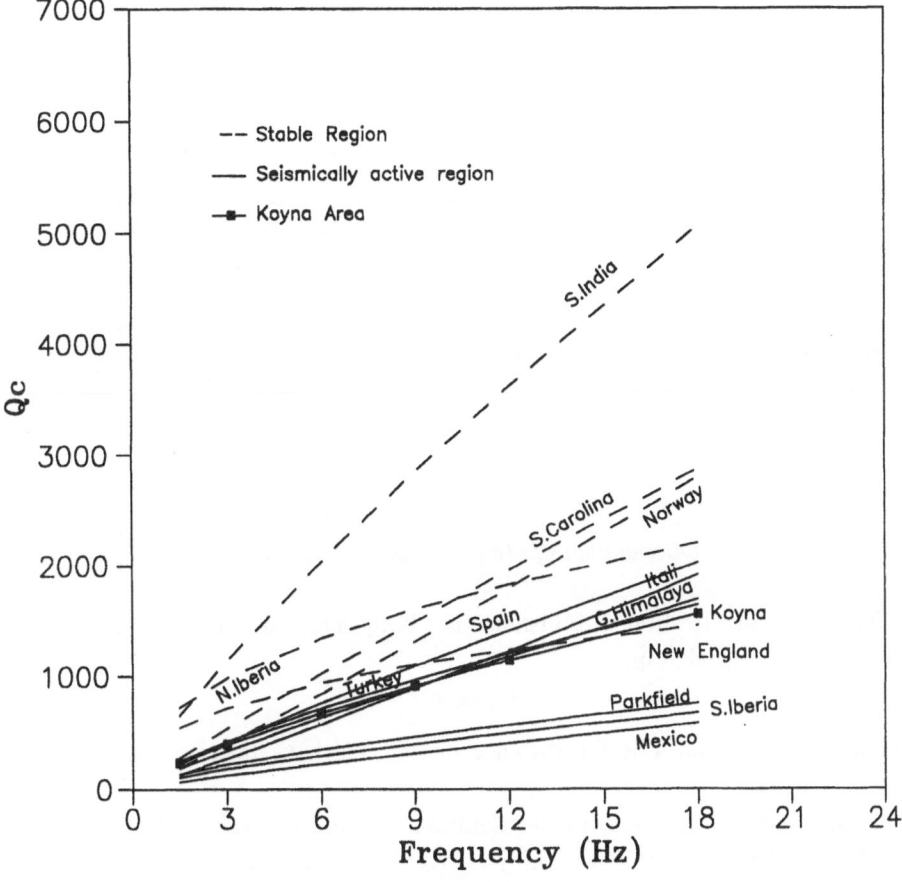

Figure 5

Comparison of Coda-Q_c estimation of the Koyna seismic zone with the reported Coda-Q_c values for seismic (solid lines) and aseismic (broken lines) regions around the world.

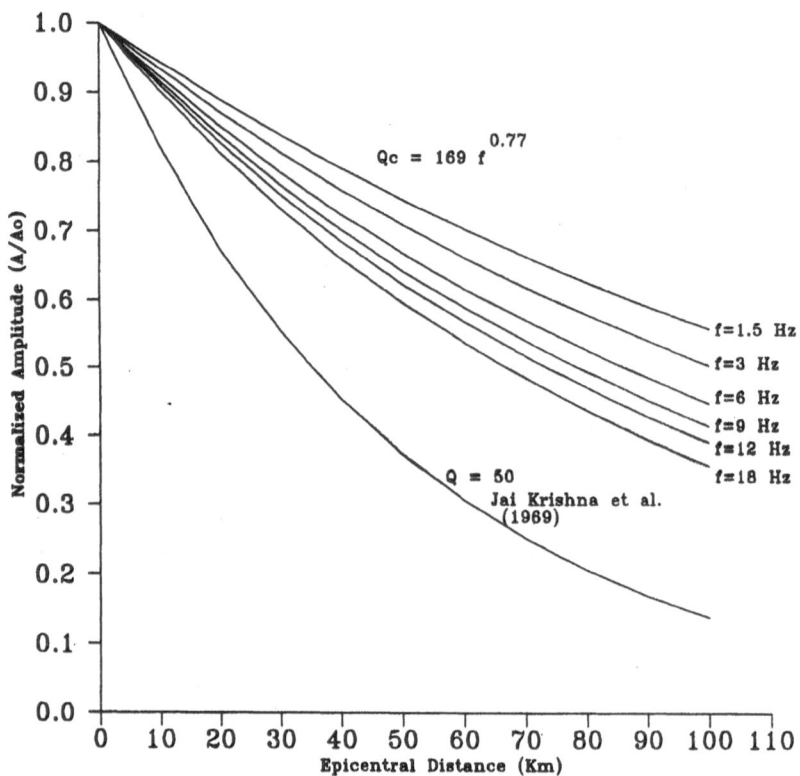

Figure 6
Decay of normalized amplitude with distance obtained for Q_c values for different frequencies from this study. The decay of normalized acceleration with distance estimated by JAI KRISHNA (1969) for the 1967 Koyna earthquake.

Attenuation of Acceleration with Distance

The fall of acceleration with distance has been obtained at different frequencies (Fig. 6) using the relation,

$$A = A_o \, e^{-\alpha x} \qquad (6)$$

where $\alpha = \pi f / \beta Q$ and x is the distance in km. We have taken S-wave velocity, $\beta = 3.5$ km/sec.

At a distance of 100 km, the acceleration is found to decrease to 60–40% for 1.5–18 Hz. This decay is considerably slower than that estimated by JAI KRISHNA et al. (1969) from the observation of damage due to the 1967 Koyna earthquake which indicated the fall to 30% at 100 km corresponding to the Q of about 50.

Comparison of our Results with Other Studies

Many of the results of Q studies obtained by various authors worldwide are shown in Figure 5 and listed in Table 1. Different regions can be broadly divided into two groups viz, active and stable.

Table 1 and Figure 5 show a low Q_o value of less than 200 for the seismically active regions around the world. The n value is about 1 for most of the active regions viz. Mexico, S. Iberia and Parkfield (as shown in Fig. 5) and also in Yugoslavia, Hindukush, Washington State, W. Africa, Dead Sea and Aleutian (not shown in Fig. 5). For Spain, Turkey, Italy and Himalaya (STIH regions) the n

Table 1

Worldwide Q studies in active and stable areas

Places	Q_o	n	Source
A: From coda-Q_c method			
Active Regions			
Guerrero, Mexico	47	0.87	RODRIGUEZ *et al.* (1983)
Yugoslavia	50	1.0	ROVELLI (1984)
Hindukush	60	1.0	ROECKER *et al.* (1982)
Washington State	63	0.97	HAVSKOV *et al.* (1989)
Mt. Cameroon, W. Africa	65	1.0	AMBEH and FAIRHEAD (1989)
Dead Sea	65	1.05	VAN ECK (1988)
Parkfield	79	0.74	HELLWEG *et al.* (1995)
Friuli, Italy	80	1.10	ROVELLI (1982)
South Iberia	100	0.70	PUJADES *et al.* (1991)
Garhwal, Himalaya	126	0.90	GUPTA *et al.* (1995)
South Spain	155	0.89	IBANEZ *et al.* (1990)
Koyna	169	0.77	This Study
West Anatolia, Turkey	183	0.76	AKINCI *et al.* (1994)
Aleutian	200	1.05	SCHERBAUM and KISSLINGER (1985)
Stable Regions			
Norway	120	1.09	KAVAMME and HAVSKOV (1988)
South Carolina	190	0.94	RHEA (1984)
South India	460	0.83	RAMAKRISHNA RAO *et al.* (1997)
New England	460	0.40	PULLI, J. J. (1984)
North Iberia	600	0.45	PUJADES *et al.* (1991)
B: From Lg wave Spectral Amplitude method			
Active Regions			
Western U.S.	150	0.4	SINGH and HERRMANN (1983)
NW U.S.	300	0.3	SINGH and HERRMANN (1983)
Stable Regions			
NE U.S.	900	0.35	SINGH and HERRMANN (1983)
Central U.S.	1000	0.20	SINGH and HERRMANN (1983)
Canadian Shield	900	0.20	HASEGAWA (1985)

value ranges from 0.7 to 0.8. The coda-Q relation for the Koyna-Warna area is similar to that for these regions with a Q_o of 169 and n of 0.77.

In stable regions the Q_o has been estimated to be generally around 500 with n about 1. However, S. Carolina and Norway have shown Q_o value of 190 and 120, respectively, which are identical to that for active regions. However, the low Q_o is limited to the low frequency range (up to 4 Hz), indicating a higher degree of heterogeneity in the upper crust. In contrast to this, N. Iberia and New England regions have shown high Q_o of 600 and 460, respectively, although due to low n value ranging from 0.4 to 0.45, the Q_c value at higher frequency range is lower than in other stable regions, rather close to the active STIH regions, indicating a relatively more attenuative lower lithosphere.

The Koyna-Warna area, though it is a part of the stable region of India, it ($Q_o = 169$ and $n = 0.77$) behaves in a different fashion than that for S. India ($Q_o = 460$ and $n = 0.83$). The attenuation property of the Koyna-Warna area is rather close to the active STIH regions.

From the above comparison of the worldwide Q studies, it can be inferred that in general the Q is low in active regions in all frequency ranges as compared to the stable regions, indicating that the entire lithosphere beneath active regions is relatively more attenuative than that of the stable regions.

Discussion

The Koyna-Warna area, which is geologically in a stable continental region, exhibits a value of Q_c similar to that of active regions. Hence, it is possible that the western Ghat region around Koyna is tectonically active and heterogeneous. The frequency dependence of attenuation indicates that the area becomes relatively less heterogeneous with depth. This situation is expected in the Deccan Trap region where the lava ascended through a few conduits and erupted through several fissures.

The frequency-dependent coda-Q_c relation obtained for the Koyna region holds good for a circular coda generation area of 40,000 km^2 with a radius of 114 km. The values obtained for the Koyna region ($Q_o = 169$ and $n = 0.77$) suggest a high n value which yields a low Q in the low frequency range (up to 4 Hz) and a high Q in the higher frequency range. The relatively low Q_c at lower frequencies (1–4 Hz) can be attributed to the loss of energy due to the presence of numerous heterogeneities and a decrease in rock strength. The large Q_c at higher frequencies may be related to the propagation of backscattered body waves in deeper parts of the upper crust where less heterogeneity is expected.

It has been observed from the comparison of worldwide Q studies that in general the Q is low in active regions in all frequency ranges as compared to the stable regions, indicating that the crust beneath active regions is relatively more attenuative than that of the stable regions.

The Q_o value estimated from the decrease in damage due to large earthquakes (e.g., about 50 for Koyna and 25 for Parkfield, as inferred from the decay curves shown by JAI KRISHNA et al., 1969) is much less than that found from the coda method (169 for Koyna, 79 for Parkfield). The Q_o estimated from the rate of fall of the recorded acceleration with distance was found to be about 50 for Uttarkashi, 1991 earthquake (as inferred from the decay curve shown by *Chandrasekaran* and DAS, 1992) which is less than the value of 126 obtained by the coda method. Hence, it may be possible that the coda method overestimates the Q, responding mostly to intrinsic attenuation and only slightly to the scattering Q_s (GAO et al., 1983). Hence, the multiple scattering technique could provide a better estimate of Q for the region.

Conclusion

This study reveals that the coda-Q_c for the Koyna-Warna seismic region is frequency dependent and defines a relation as $Q_c = 169 \pm 3.4 f^{0.77 \pm 0.14}$. The estimated Q_c ($=169$) and n ($=0.77$) values are like that for several seismically active regions of the world, though the area lies in the stable continental region of India. It indicates that the lithosphere beneath the Koyna-Warna region is quite heterogeneous. The Q_c value represents the average attenuation property for a circular area of over 40,000 km^2 (radius of 114 km). The Q_c values are found to range from 169 at 1 Hz to 1565 at 18 Hz. The low Q_c values at lower frequency range (1–3 Hz) can be attributed to the energy loss due to the presence of numerous heterogeneities. The high Q values for the higher frequency band (≥ 12 Hz) may be related to the relatively more homogeneous deeper crustal layers. With a dense network of digital stations, an iso-Q map is planned for the area. The Q_c estimated for larger magnitude earthquakes, using different techniques, might yield better results.

Acknowledgements

The authors are thankful to Dr. H. K. Gupta, Director, NGRI for his encouragement and kind permission to publish this work. This study was supported by the Department of Science and Technology, New Delhi. The Maharashtra Engineering Research Institute, Nasik readily provided their data and infrastructure facilities.

REFERENCES

AKI, K., and CHOUET, B. (1975), *Origin of Coda Waves: Source, Attenuation and Scattering Effects*, J. Geophys. Res. *80*, 3322–3342.

AKINCI, A., TAKTAK, A. G., and ERGINTAV, S. (1994), *Attenuation of Coda Waves in Western Anatolia*, Phys. Earth Planet. Int. *87*, 155–165.

AMBEH, W. B., and FAIRHEAD, J. D. (1989), *Coda-Q Estimates in the Mount Cameroon Volcanic Region, West Africa*, Bull. Seismol. Soc. Am. *79*, 1589–1600.

CHADHA., R. K., GUPTA, H. K., KUMPELL, H.J., MANDAL, P., NAGESWARA RAO, A., NARENDRA KUMAR, RADHAKRISHNA, I., RASTOGI, B. K., RAJU, I. P., SARMA, C. S. P., SATYAMURTHY, C., and SATYANARAYANA, H. V. S. (1997), *Delineation of Active Faults, Nucleation Process and Pore Pressure Measurements at Koyna (India)*, Pure appl. geophys. 150, 551–562.

CHANDRASEKARAN, A. R., and DAS, J. S. (1992), *Analysis of Strong Motion Accelerograms of Uttarkashi Earthquake of October 20, 1991*, Bull. Indian Soc. Earthq. Technol. *29* (1), 35–55.

GAO, L. S., LEE, L. C., BISWAS, N. N., and AKI, K. (1983), *Comparison of the Effects between Single and Multiple Scattering on Coda Waves for Local Earthquakes*, Bull. Seismol. Soc. Am. *73*, 377–389.

FRANKEL, A. D., and WENNERBERG, L. (1987), *Energy-flux Model of Seismic Coda: Separation of Scattering and Intrinsic Attenuation*, Bull. Seismol. Soc. Am. *77*, 1223–1251.

GUPTA, S. C., SINGH, V. N., and ASHWANI KUMAR (1995), *Attenuation of Coda Waves in the Garhwal Himalaya, India*, Phys. Earth Planet. Int. *87*, 247–253.

GUPTA, H. K., RASTOGI, B. K., and NARAIN, H. (1972a), *Common Features of the Reservoir Associated Seismic Activities*, Bull. Seismol. Soc. Am. *62*, 481–492.

GUPTA, H. K., RASTOGI, B. K., and NARAIN, H. (1972b), *Some Discriminatory Characteristics of Earthquakes near the Kariba, Kremasta and Koyna Artificial Lakes*, Bull. Seismol. Soc. Am. *62*, 493–507.

GUPTA, H. K., RAO, C. V. R. K., and RASTOGI, B. K. (1980), *An Investigation of Earthquakes in Koyna Region, Maharashtra, for Period October 1973 through December 1976*, Bull. Seismol. Soc. Am. *70*, 1833–1847.

HASEGAWA, H. S. (1985), *Attenuation of Lg Waves in the Canadian Shield*, Bull. Seismol. Soc. Am. *75*, 1569–1582.

HAVSKOV, J., MALONE, S., McCLURY, D., and CROSSON, R. (1989), *Coda-Q for the State of Washington*, Bull. Seismol. Soc. Am. *79*, 1024–1038.

HELLWEG, M., SPANDICH, P., FLETCHER, J. B., and BAKER, L. M. (1995), *Stability of Coda Q in the Region of Parkfield, California: View from the U.S. Geological Survey Parkfield Dense Seismograph Array*, J.G.R. *100*, 2089–2102.

IBANEZ, J. M., DEL PEZZO, E., DE MIGUEL, F., HERRAIZ, M., ALGUACIL, G., and MORALES, J. (1990), *Depth Dependent Seismic Attenuation in the Granada Zone (Southern Spain)*, Bull. Seismol. Soc. Am. *80*, 1222–1234.

JAI KRISHNA, CHANDRASEKARAN, A. R., and SAINI, S. S. (1969), *Analysis of Koyna Accelerograms of December 11, 1967*, Bull. Seismol. Soc. Am. *59*, 1719–1731.

JIN, A., and AKI, K. (1988), *Spatial and Temporal Correlation between Coda Q and Seismicity in China*, Bull. Seismol. Soc. Am. *78*, 741–769.

KAVAMME, L. B., and HAVSKOV, J. (1988), *Q in Southern Norway*, Bull. Seismol. Soc. Am. *79*, 1575–1588.

PUJADES, L., CANAS, J.A., EGOZCUE, J. J., PUIGVI, M. A., POUS, J., GALLART, J., LANA, X., and CASAS, A. (1991), *Coda Q Distribution in the Iberian Peninsula*, Geophys. J. Int. *100*, 285–301.

PULLI, J. J. (1984), *Attenuation of Coda Waves in New England*. Bull. Seismol. Soc. Am. *74*, 1149–1166.

RAMAKRISHNA RAO, C. V., SESHAMMA, N. V., and MANDAL, P. (1997), *Estimation of Coda Q_c and Spectral Characteristics of Some Moderate Earthquakes of Southern Indian Peninsula* (communicated to Bull. Seismol. Soc. Am.).

RASTOGI, B. K. (1992), *Seismotectonics Inferred from Earthquakes and Earthquake Sequences in India during the 1980s*, Curr. Sci. *62*, 101–108.

RASTOGI, B. K., and TALWANI, P. (1980), *Relocation of Koyna Earthquakes*. Bull. Seismol. Soc. Am. *70*, 1849–1868.

RASTOGI, B. K., CHADHA, R. K., SARMA, C. S. P., MANDAL, P., SATYANARAYANA, H .V. S., NARENDRA KUMAR, RAJU, I. P., SATYAMURTHY, C., and NAGESWARA RAO, A. (1997), *Seismicity at Warna Reservoir (near Koyna) through 1995*, Bull. Seismol. Soc. Am. *87*, 1484–1494.

RAUTIAN, T. G., and KHALTURIN, V. I. (1978), *The Use of the Coda for Determination of the Earthquake Source Spectrum*, Bull. Seismol. Soc. Am. *68*, 923–948.

RHEA, S. (1984), *Q Determined from Local Earthquakes in the South Carolina Coastal Plain*, Bull. Seismol. Soc. Am. *74*, 2257–2268.

RODRIGUEZ, M., HAVSKOV, J., and SINGH, S. K. (1983), *Q from Coda Waves near Petatlan, Guerrero, Mexico*, Bull. Seismol. Soc. Am. *73*, 321–362.

ROECKER, S. W., TUCKER, B., KING, J., and HARTZFELD, D. (1982), *Estimates of Q in Central Asia as a Function of Frequency and Depth Using the Coda of Locally Recorded Earthquakes*, Bull. Seismol. Soc. Am. *72*, 129–149.

ROVELLI, A. (1982), *On the Frequency Dependence of Q in Friuli from Short Period Digital Records*, Bull. Seismol. Soc. Am. *72*, 2369–2372.

ROVELLI, A. (1984), *Seismic Q for the Lithosphere of the Montenegro Region (Yugoslavia): Frequency, Depth and time Windowing Effects*, Phys. Earth Planet. Inter. *34*, 159–172.

SATO, H. (1977), *Energy Propagation Including Scattering Effects, Single Isotropic Approximation*, J. Phys. Earth. *25*, 27–41.

SATO, H. (1988), *Fractal Interpretation of the Linear Relation between Logarithms of Maximum Amplitude and Hypocentral Distance*, Geophys. Res. Lett. *15*, 373–375.

SATO, H. (1994), *Multiple Isotropic Scattering Model Including P-S Conversions for the Seismogram Envelope Formation*, Geophys. J. Int. *117*, 487–494.

SCHERBAUM, F., and KISSLINGER, C. (1985), *Coda Q in the Adak Seismic Zone*, Bull. Seismol. Soc. Am. *75*, 615–620.

SINGH, S., and HERRMANN, R. B. (1983), *Regionalization of Crustal Coda Q in the Continental United States*, J. Geophys. Res. *88*, 527–538.

SYKES, L. R. (1970), *Seismicity of the Indian Ocean and a Possible Nascent Island Arc between Ceylon and Australia*, J. Geophys. Res. *75*, 5041–5055.

TANDON, A. N., and CHAUDHURY, H. M. (1969), *Koyna Earthquake of December, 1967*, India Meteorol. Dep., Sci. Rep. *59*, 12 pp.

TOPPING, J., *Errors of Observation and their Treatment* (Chapman and Hall Ltd., London 1963), pp. 105–109.

VAN ECK, T. (1988), *Attenuation of Coda Waves in the Dead Sea Region*, Bull. Seismol. Soc. Am. *2*, 770–779.

WU, R. S., and AKI, K. (1985), *Elastic Wave Scattering by a Random Medium and the Small-scale Inhomogeneities in the Lithosphere*, J. Geophy. Res. *90*, 10261–10273.

ZENG, Y. (1991), *Compact Solutions for Multiple Scattered Wave Energy in the Time Domain*, Bull. Seismol. Soc. Am. *81*, 1022–1029.

(Received, July 9, 1997, revised/accepted March 10, 1998)

Pure appl. geophys. 153 (1998) 179–194
0033–4553/98/010179–16 $ 1.50 + 0.20/0

Pure and Applied Geophysics

Aquifer-induced Seismicity in the Central Apennines (Italy)

F. Bella,[1] P. F. Biagi,[2] M. Caputo,[3] E. Cozzi,[1] G. Della Monica,[1]
A. Ermini,[4] W. Plastino,[1] and V. Sgrigna[1]

Abstract—The Gran Sasso chain (Central Apennines, Italy) contains one of the largest aquifers of Central Italy. From 1970–1986 the massif was tunnelled through in order to build up a highway and an international underground laboratory for nuclear physics research. These works have strongly modified the hydrogeological situation of the chain, as shown by the decrease in flow rate that occurred in many springs located at the border of the carbonatic structure, along the boundary between the permeable limestone of the massif and the surrounding aquicludes. The analysis of the seismicity ($M \geq 3.0$) that occurred in the Gran Sasso area from 1956 to 1995 suggests that after the tunnelling works both the number of earthquakes has increased and epicenters have migrated, gathering at the northwestern border zone. The foremost events which occurred in this zone in recent years took place on May 5, 1992 ($M = 3.1$), August 25, 1992 ($M = 3.9$) and March 13, 1994 ($M = 3.5$). The flow rate data of four springs and water level data of an underground karst pool located at the border of the carbonatic structure of the massif show clear anomalies before the occurrence of the quoted earthquakes. Regardless, these anomalies can be explained by the rapid melting of the thick mantle of snow on the Gran Sasso chain, due to sudden increases of mean temperatures. In this paper we present and discuss the possibility that the quoted earthquakes are induced by the irregular variations of the Gran Sasso aquifer, evidenced by the quoted anomalies in the flow rate and water level.

Key words: Induced seismicity, flow rate, water level, preseismic anomalies, aquifer.

1. Introduction

Induced seismicity is a phenomenon generally connected with reservoirs. Experimental results were obtained for Nurek reservoir (SIMPSON and NEGMATULLAEV, 1981; KEITH et al., 1982), Lake Mead (ANDERSON and LANEY, 1975; ROGERS and LEE, 1976), Lake Kariba (GOUGH and GOUGH, 1976; SNOW, 1982), Lake Oroville (BELL and NUR, 1978). Two considerations result from them:
a) the first reservoir impoundment can lead to seismicity in a quiescent area (SIMPSON and NEGMUTALLAEV, 1981; KEITH et al., 1982);

[1] Dipartimento di Fisica, Università di Roma III, Via della Vasca Navale, 84-00146 Roma, Italy.
[2] Dipartimento di Fisica, Università di Bari, Via Amendola, 173-70126 Bari, Italy.
[3] Dipartimento di Fisica, Università "La Sapienza," Piazzale A. Moro 2, 00185 Roma, Italy.
[4] Dipartimento di Scienze e Tecnologie Fisiche ed Energetiche, Università "Tor Vergata," Via di Tor Vergata, 00133 Roma, Italy.

b) successive cyclic variations in the water level of a reservoir produce induced
   seismicity for several years after the first impoundment (BELL and NUR, 1978;
   ROELOFFS, 1988).
This seismicity can be triggered on pre-existing faults subject to near critical
tectonic stress, located both beneath the reservoir and the border zone. The trigger
seems to be caused by a reduction in fault stability S, given by

$$\Delta S = \mu(-\Delta\sigma - \Delta p) - \Delta\tau \tag{1}$$

where μ is the coefficient of friction, τ is shear stress on the fault in the direction
of slip, $-\sigma$ is the compressive normal stress across the fault, and p is pore pressure
(ROELOFFS, 1988). The calculated reduction in S is at most a few tenths of a
megapascal.

Some earthquakes occurring in zones where no reservoirs exist could be
considered induced earthquakes if some subsurface hydrodynamic changes which
are able to destabilize faults can be found.

One of the largest aquifers in Central Italy is located in the Gran Sasso chain
(see Section 2). From 1970 to 1980 this massif was completely tunnelled through, in
order to construct a highway (ANAS-COGEFAR, 1980). During the period from
1982 to 1986 other excavations were carried out for the construction of the INFN
(National Institute of Nuclear Physics) international underground laboratory. Since
1976, drainage connected with the tunnelling resulted in a remarkable decrease of
the flow rate in the northern springs. In some cases this decrease reached 70%. The
water supply for some villages located around the chain was reduced. Two drainage
springs were realized at both entrances of the tunnel but the overall hydraulic
balance is a loss. Consequently, the excavation works provoked strong hydrologic
changes in the chain. The subsequent condition is thus very similar to that
occurring during the first impoundment of a reservoir and induced seismicity can
therefore take place. In this case, a detailed study of induced seismicity presents
three main difficulties:

a) the shape and size of an aquifer are generally little known with respect to a
   reservoir;
b) the faulting beneath and at the border zone of an aquifer are unknown, although
   this does not happen for a reservoir where many preliminary geological investigations are carried out with great accuracy;
c) seismicity that occurs close to an aquifer is generally known with little precision,
   especially for small magnitudes, while local seismic networks always located
   around a reservoir can provide considerable precise seismic data.

In spite of these difficulties, we investigated both the seismicity in the Gran Sasso
area and several hydrologic data sets and in this paper we present the possibility
that the three strongest ($M = 3.1-3.9$) earthquakes which occurred in the Gran
Sasso northwestern area in recent years can be considered induced earthquakes.

2. The Gran Sasso Chain

The Gran Sasso chain (Fig. 1) is a tectonic unit formed by a double chain separated by deep trenches and high plains and with deep depressions at the boundaries. The northern chain includes some of the highest elevations in the Central Apennines (2400–3000 m), while the southern chain has lower altitudes (1600–2000 m). There is no drainage system, and karst features are extensively distributed inside the massif due to the mostly complete absorption of waters and the presence of limestone.

The Gran Sasso chain is formed mainly by carbonatic units which thrust northwardly over clay-marly sandstones (Fig. 2). The southern slope consists of a series of ridges stretching out along a NW–SE direction, and sloping down southwards. These ridges extend under alluvial deposits of the Aterno River Valley (GHISETTI, 1987; GHISETTI and VEZZANI, 1991).

A detailed study of this structure (ANAS-COGEFAR, 1980) was performed by drilling three very deep bore holes before the tunnel construction and establishing the INFN underground laboratory (1970–1985).

The Gran Sasso chain is the northern part of the wide hydrogeological system (BONI et al., 1986) formed by the Gran Sasso and the Velino-Sirente mountains (Fig. 1). The total area of this complex is about 2164 km^2 and it can absorb 714 mm of annual effective infiltration and can give back 49 m^3/s as mean discharge. Inside the chain a very large and deep karst aquifer is located. It consists of a series of minor aquifers with decreasing altitudes along a northeast–southwest direction (Fig. 2a), separated by the main structural discontinuities, but connected one with the other, so that the water-bearing stratum at the bottom can be considered a unique, large aquifer. From the three above-mentioned borings, values of the permeability k not higher than 10^{-9} cm^2 have been obtained. In karst zones k values increase considerably (ANAS-COGEFAR, 1980). This variability of permeability produces different water levels in the aquifer. Springs are located at the border of the carbonatic structure, along the boundary between permeable limestone and aquicludes. In particular, the arenaceous flysch complex of the Laga Mountains (Figs. 1 and 2b) acts as an aquiclude at the northern slope of the Gran Sasso chain and displays k values of the order of 10^{-11} cm^2. Due to altitude, seasonal variations of the aquifer occur mainly during the thaw and different meteorological conditions above the northern and the southern slopes of the chain provoke different recharges of the aquifer. The largest variations of water level are of the order of several tens of meters (ANAS-COGEFAR, 1980). In any case, it seems reasonable to assume that the changes in the aquifers of the chain are reflected by the changes in flow rates of the surrounding springs.

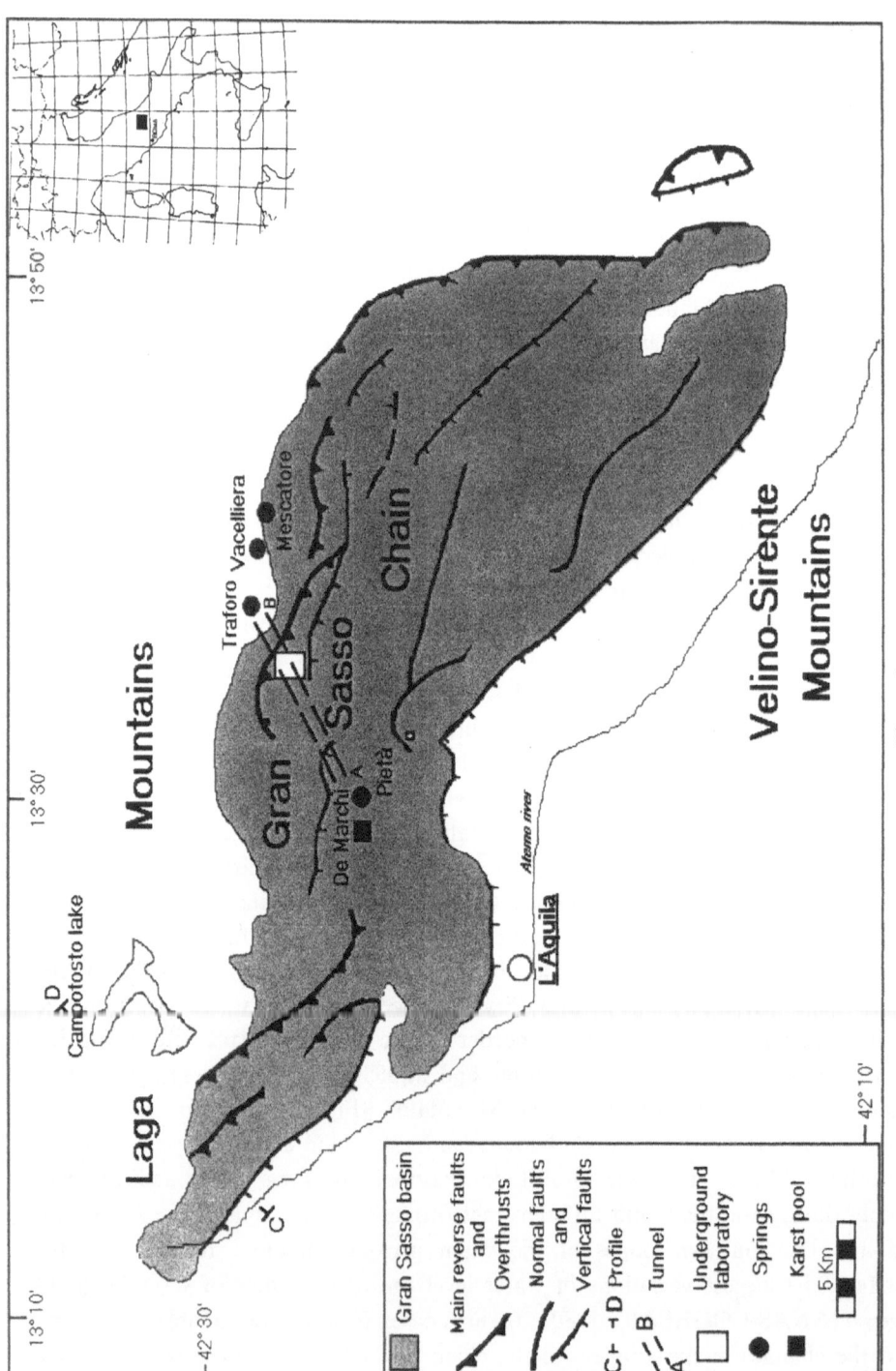

Figure 1

Map showing the Gran Sasso basin with the tunnel and the principal fault systems. The location of the underground laboratory together with the springs and the karst pool considered in this paper is also reported.

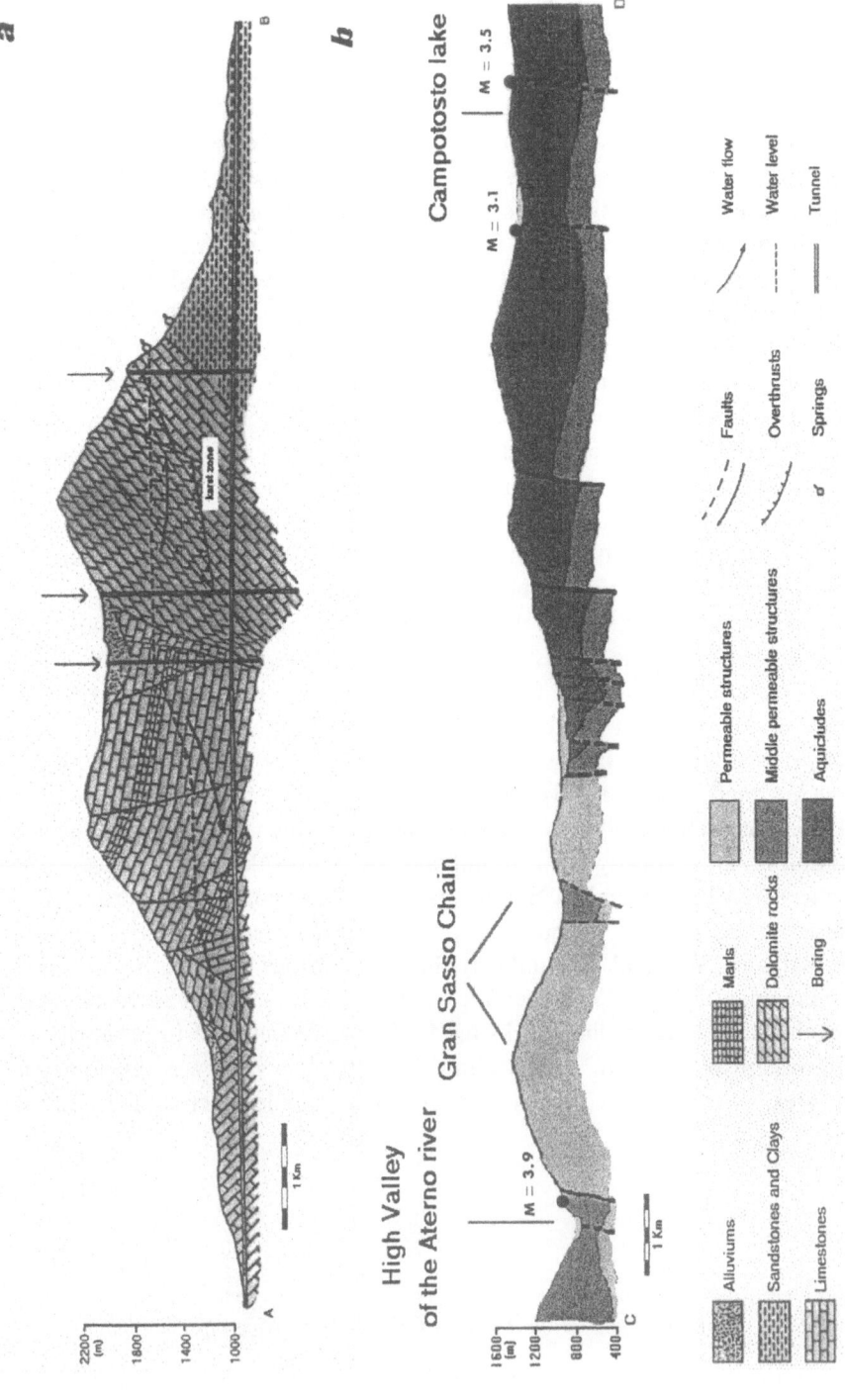

Figure 2

(a) Geologic section of the Gran Sasso chain along the tunnel direction from ANAS-COGEFAR (1980). The three holes and some springs at the northern border of the chain are shown together with the main karst zone and the water level of the aquifer. (b) Schematic hydrologic section from High Valley of the Aterno River to Campotosto Lake, displaying permeable structures and aquicludes. Dots represent the epicenters of the August 25, 1992 ($M = 3.9$), May 5, 1992 ($M = 3.1$) and March 13, 1994 ($M = 3.5$) earthquakes.

3. Seismicity Pattern

Seismicity in the Gran Sasso area has been studied by means of the bulletins of the ING (National Institute of Geophysics). From 1975 to 1985 the Italian seismic network was modernized and the number of stations increased. Therefore, the completeness of seismic catalogs varies with time. If earthquakes with $M \geq 3.0$ are considered, it is not unlikely that the semismic catalog concerning the quoted area is quite homogeneous, starting from 1956, when L'Aquila seismic station began operation. The area has been split into two different zones named "inside" and "outside". The inside zone represents the hydrogeological basin of the Gran Sasso chain; the outside zone is a perimetrical belt 10 km wide, surrounding the inside zone. This division is similar to those used in seismic studies concerning reservoirs. The annual frequency of earthquakes with $M \geq 3.0$ which occurred each year from 1956 until 1995 in these two zones is shown in Figure 3, together with their energy release. The spatial seismicity divided into ten-year periods is also shown in Figure 3. The largest energy release occurred in 1958 and it was related to an earthquake that took place on June 24, 1958 with $M = 5.0$ in a zone to the south of L'Aquila city. This event is representative of the tectonics that are now weakly active in this area, although it was violent from 1300 to 1700 when catastrophic earthquakes occurred. On the basis of the results shown in Figure 3, we can make the following further observations. Initially, in the period under examination, seismicity with $M \geq 3.0$ occurred mainly in the border zone. Next, the number of earthquakes ($M \geq 3.0$) per year increased in this zone after the excavation works. Regarding the energy release, there have been two maxima during the last twenty years, in 1980 and in 1985, that is after the two successive drillings through the massif. A further consideration is that seismicity seems to migrate after the works; in fact, since 1976, the epicenters are aligned mainly along the major axis of the basin and, particularly, at the western zone, contrary to previous years. Also the energy release is concentrated in that zone. We think it is unlikely that the previous evidence is connected only with the time variability of the seismic catalog. Consequently, it seems possible to deduce that the big tunnelling works inside the massif have led to seismicity in the north-western border zone of the Gran Sasso basin, which was previously a quiescent area, even if enclosed in an active tectonic complex such as the Central Apennines.

4. Hydrology and Seismicity

4.1. Hydrologic Anomalies and High Valley of the Aterno River Seismicity

The flow rate data of Vacelliera, Traforo, Mescatore and Pietà springs (Fig. 1) have been analyzed. Since 1986, the flow rate in the springs has been sampled once

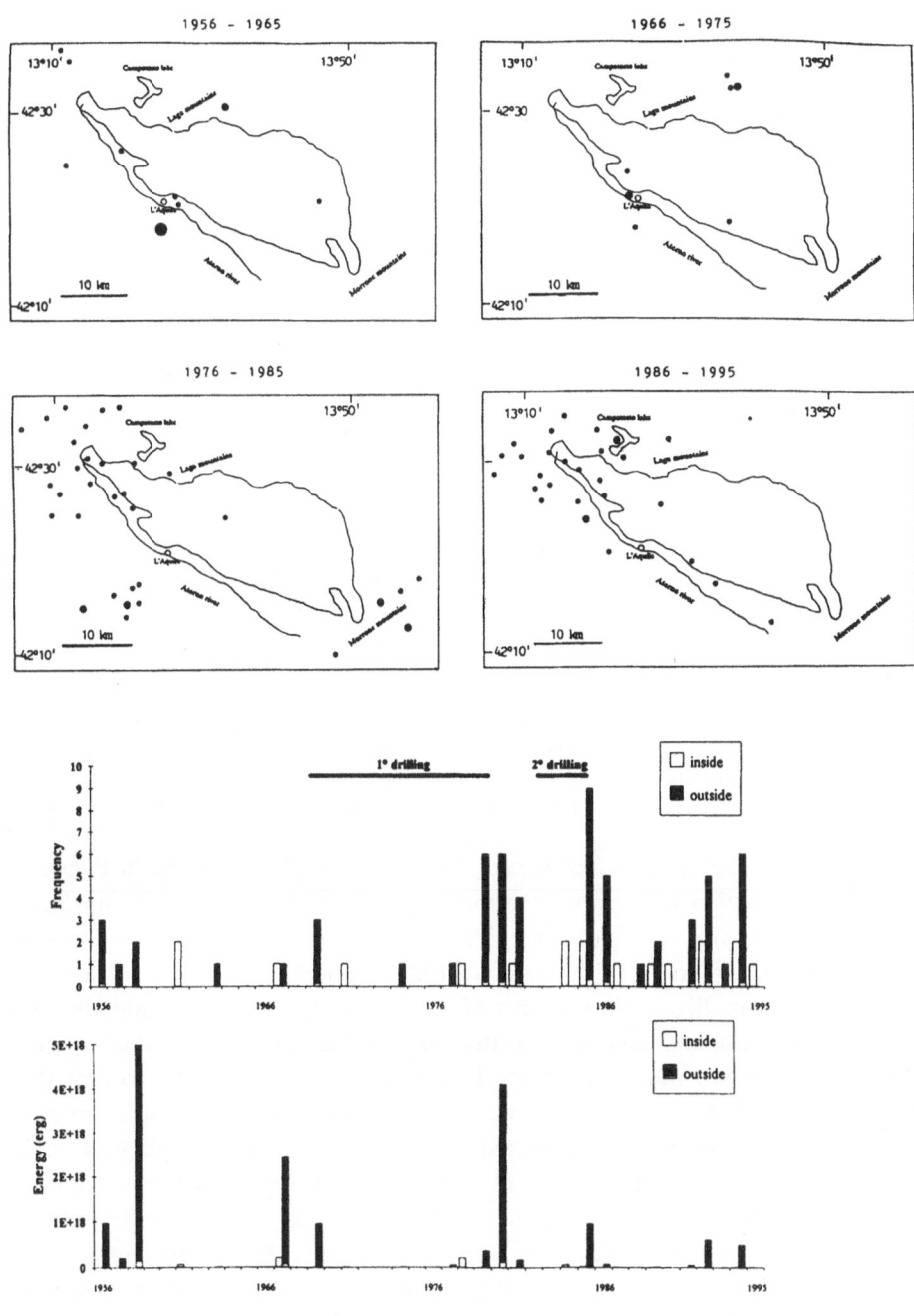

Figure 3

The spatial distribution of the seismicity which occurred in the last four ten-year periods is shown in the squares ($\cdot$ $3.0 \leq M < 3.5$, $\bullet$ $3.5 \leq M < 4.0$, $\bullet$ $M \geq 4.0$). The annual variation of seismicity (frequency and energy) which occurred inside and outside the basin from 1956 to 1995 is reported at the bottom. The time intervals for the two drillings through the Gran Sasso chain are indicated.

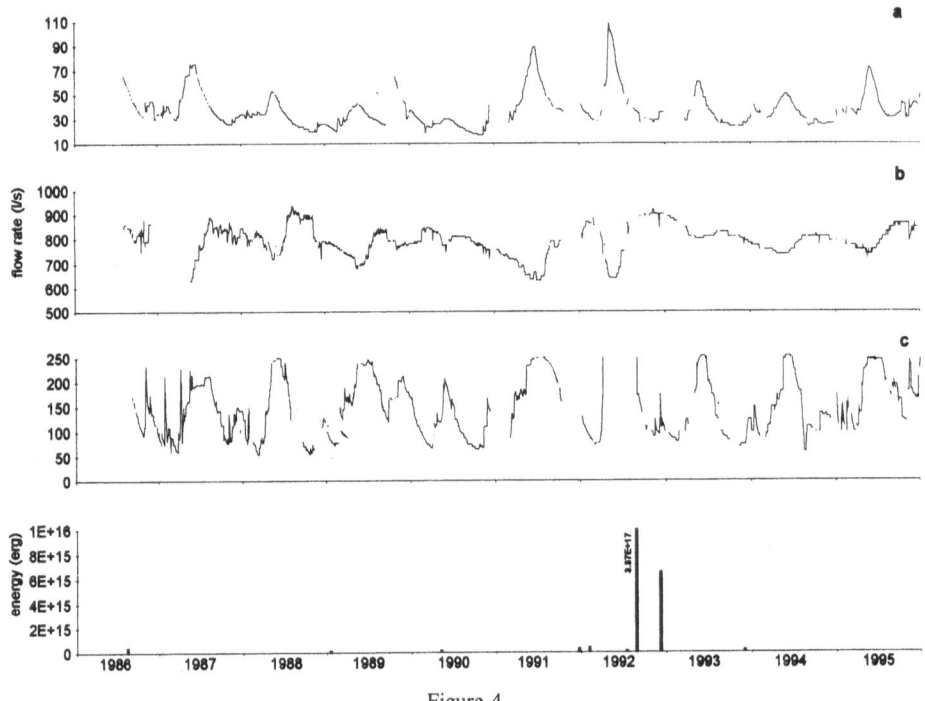

Figure 4

Flow rate trends of Vacelliera (a), Traforo (b), and Mescatore (c) springs from 1986 to 1995. Interruptions in the trends are related to lack of data; from the beginning of April to mid August 1992 data of Mescatore spring are over scale. At the bottom, the energy release of the earthquakes which occurred in the seismogenetic area of the High Valley of the Aterno River from 1986 to 1995 is shown.

a day by the Ruzzo Waterworks Society. Flow rate trends are shown in Figure 4. The annual variation linked with the discharge and recharge of the Gran Sasso aquifer is evident from the plot. The phase-reversal trend of the Traforo spring is coupled with the nature of its drainage. In 1992 an anomaly in the cyclic variation appears clearly in Figure 4. Features of this anomaly are both amplitude and rapidity in increase and decrease. On the basis of data provided by the National Meteorologic Service and the L'Aquila Tourist Board, it was possible to note that this irregularity is associated with particular seasonal meterological conditions on a regional scale, consisting of abrupt temperature increases that result in a rapid melting of the thick mantle of snow (2 or 3 meters) on the Gran Sasso chain.

The seismicity which occurred in the northwestern area of the Gran Sasso basin during the experimental period has been studied purposeful to finding possible correlations with the flow rate anomaly. During these last ten years local seismic networks in the area were established (National Seismic Service-University of L'Aquila, ENEL-ISMES) which facilitated the possibility of considering earthquakes with a minima magnitude of 2.0. The energy release of the earthquakes which occurred in the seismogenetic area of the High Valley of the Aterno River

(Fig. 1) from 1986 to 1995 is shown at the bottom of Figure 4. The epicenters are reported in Figure 5a. The most remarkable event was a seismic sequence that occurred on August 25, 1992, characterized by an earthquake with $M = 3.9$ followed by several aftershocks with $M \leq 2.0$ in the following days. The focal mechanism of this event is shown in Figure 5a. The estimation of source parameters of the seismic sequence (BONGIOVANNI et al., 1993) gives rather low stress drop values (0.01–1.5 MPa) with respect to those generally associated with tectonic earthquakes along normal faults.

Figure 4 demonstrates clearly that the irregular variation of the flow rates in 1992 is an anomaly related to the quoted seismic sequence. The interpretation of this anomaly in a tectonic sense would suggest that tectonic stresses can surface in coincidence with a particular meteorological condition such as that mentioned above and this circumstance seems unlikely. However if we consider the anomalous variation of the Gran Sasso aquifer in 1992 as a trigger of the following seismic sequence, this perplexity disappears. Therefore, the main shock of the sequence can be numbered among the induced earthquakes.

4.2. Hydrologic Anomalies and Campotosto Lake Seismicity

The flow rate data of Pietà spring and level data of the De Marchi karst pool (Fig. 1) have been analyzed. Since 1992, we have measured the flow rate of the Pietà spring every fifteen days. The De Marchi pool is a little karst pool, 10 meters in diameter, placed at the bottom of a cave (Amare cave) near the Pietà spring. Dives exceeding 30 meters were unable to reveal the bottom of this pool; its groundwater level is in agreement with the altitude of some springs. Thus, it is very probable that this pool is linked with minor aquifers of the whole basin (see previous section). Since 1985 our group has been carrying out research inside the Amare cave (BELLA et al., 1994, 1995). At the bottom of the cave an undergroumnd laboratory was set up for geophysical studies, in particular measurements of electric, magnetic and seismoacoustic fields. In 1991 we installed a level meter inside the De Marchi pool which measures the variation in level every ten minutes.

Figure 6 illustrates the flow rate trend of the Pietà spring and the level trend of the De Marchi pool from January 1992 to December 1995. The smoothed trend obtained using a high period pass filter ($T = 6$ months) is represented by the dotted line in Figure 6 and it is in agreement with the raw flow rate trends we presented previously. It can therefore be considered as the regular representation of discharge and recharge of the Gran Sasso aquifer. From this point of view the status is regular in 1993 as concerns the Pietà spring flow rate and the De Marchi pool level; otherwise, irregularities are evident in both trends during the other years, particularly in 1992 and 1994, when the increase rate of the pool level was evaluated as about 6 cm/d. In 1995 the increase rate was about 3 cm/d. The irregularities pointed out can be considered a peculiarity of the minor aquifer with which they are

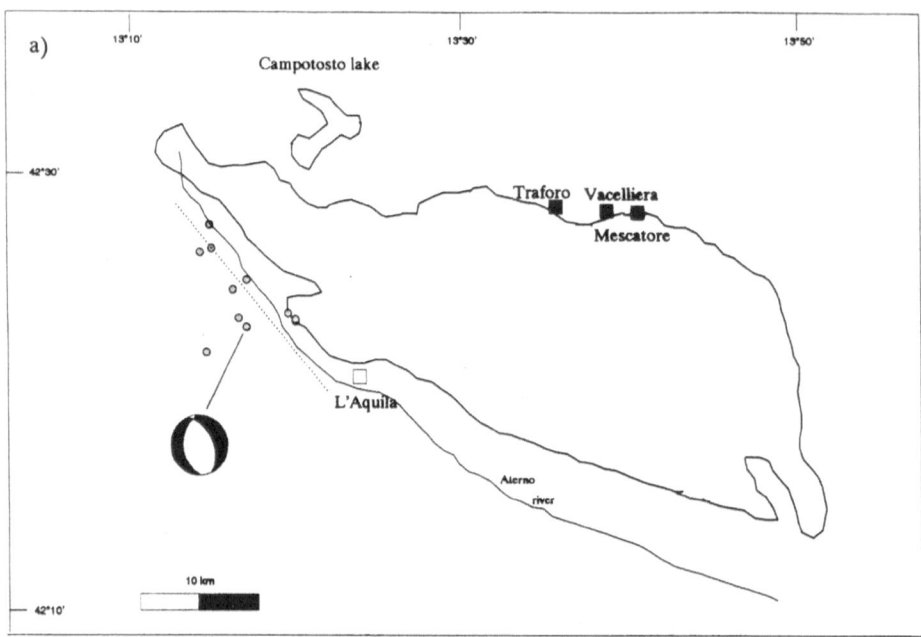

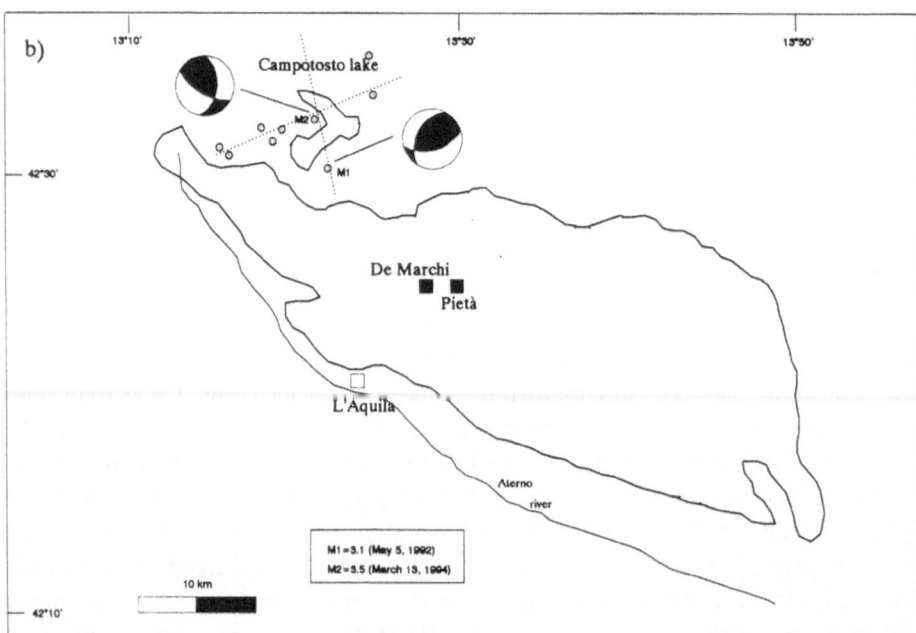

Figure 5

Map showing the epicenters of the earthquakes which occurred: a) in the area of the High Valley of the Aterno River from 1986 to 1995; b) in the Campotosto Lake area from 1992 to 1995. The three earthquakes with the greatest magnitude ($M > 3.0$) are indicated with their own focal mechanism. Dotted lines delineate the main fault alignments in the two zones. The location of Pietà, Vacelliera, Traforo, Mescatore springs and of the De Marchi pool is also reported.

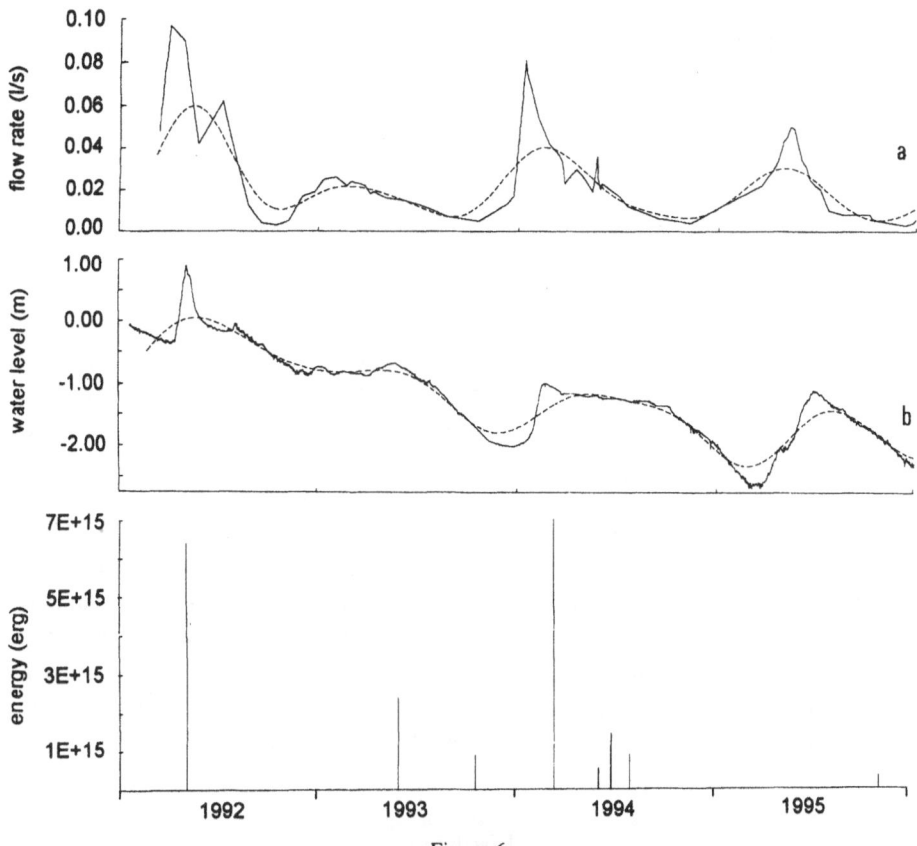

Figure 6

Flow rate trend of Pietà spring (a) and water level of De Marchi karst pool (b) from 1992 to 1995. Dotted lines represent filtered trends with high period pass filter ($T = 6$ months). At the bottom the energy release of the earthquakes which occurred in the seismogenetic area of Campotosto Lake from 1992 to 1995 is shown.

connected. From Figure 6 a long term decrease in the De Marchi pool in the water level appears. This decrease is probably a further manifestation of the hydrologic changes in the Gran Sasso aquifer subsequent to the excavation works (see previous sections). The two most evident anomalies displayed in Figure 6 can be easily joined with characteristic meteorological conditions. In fact, in 1992 the circumstance was particularly on a regional scale as mentioned; on the basis of the National Meteorologic Service data and the L'Aquila Tourist Board, in 1994 in a similar situation arose, however it particularly involved the southern slope of the massif where the Pietà spring and the De Marchi pool are located.

The seismicity which occurred in the northwestern zone of the Gran Sasso basin correlates with the examined hydrologic variations also in this case. The energy release of the earthquakes ($M \geq 2.0$) which occurred in the seismogenetic area of Campotosto Lake (Fig. 1) from 1992 to 1995 is shown at the bottom of Figure 6.

Epicenters of these earthquakes are reported in Figure 5b. The two strongest events occurred on May 5, 1992 ($M = 3.1$) and March 13, 1994 ($M = 3.5$). Their focal mechanisms are also shown in Figure 5b. Figure 6 shows clearly that the irregular variations of the De Marchi pool level and the Pietà spring flow rate in 1992 and in 1994 can be considered as anomalies related to the two earthquakes mentioned above. Therefore, by means of the same previous considerations, we can consider these earthquakes as induced by the hydrologic variations of some minor aquifer of the Gran Sasso basin. Also, in this case they can be considered amongst the induced ones.

Campotosto Lake extends for about 8 km² at the southwestern slope of the Laga Mountains (Fig. 1). It is a reservoir of an hydroelectric plant, where water flows from the Laga Mountains and, mainly, from the Gran Sasso chain. The plant was built up in the 1940s; in 1963 it was restored in order to achieve more storage. Now the mean depth is about 15 m. From 1983 to 1993 a local seismic network with five stations (ENEL-ISMES) operated in this area, with the aim of examining possible induced seismicity connected with water level fluctuations. About 800 earthquakes ($0.2 \leq M \leq 3.5$) beneath and at the edge (perimetrical belt 10 km wide) of the reservoir were recorded and no particular evidence of induced local seismicity was found (R. Berardi—ENEL, private communication). As an example, Figure 7 shows the water level trend of Campotosto Lake from 1983 to 1993 with the monthly number of earthquakes and the energy release related to the events with $M \geq 2.5$ which occurred beneath and at the edge of the reservoir.

5. Discussion

Initially we consider those earthquakes that we have considered as induced by the anomalous variations of the Gran Sasso aquifer. As regards the seismogenetic area of Campotosto Lake, the two earthquakes are located in the aquicludes of the Laga Mountains (Figs. 2b and 5b), with their hypocenters at about 8 km (May 5, 1992) and 13 km (March 13, 1994). In both cases their focal mechanisms are consistent with reverse faults. As concerns the seismogenetic area of the High Valley of the Aterno River, the main shock of the seismic sequence (Figs. 2b and 5a) is located in the carbonatic structure under the aquiclude which constitutes the Aterno Valley, and which is a southwestward extension of the structure of the Gran Sasso massif. The estimated hypocentral depth is about 9 km and the focal mechanism is consistent with normal fault. Delays between the three considered earthquakes, with respect to the hydrologic variations with which they are connected, are reported in Table 1. Onset times of the hydrologic variations are estimated on the basis of the De Marchi pool and the Vacelliera spring.

In studies concerning induced seismicity, pore pressure plays a fundamental role as a destabilizing factor in relationship (1). Pore pressure diffusion in a uniform

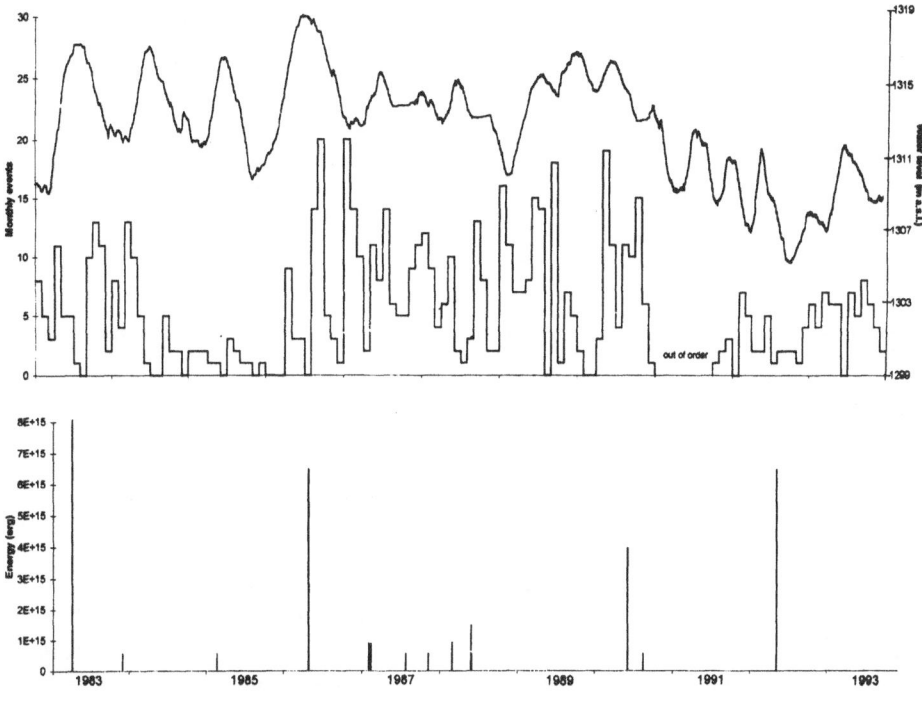

Figure 7

At the top: water level trend of Campotosto Lake from 1983 to 1993 with the monthly frequency of earthquakes which occurred within a distance of 10 km. At the bottom: energy release of the earthquakes with $M \geq 2.5$.

porous elastic medium has been studied by RICE and CLEARY (1976), BELL and NUR (1978), TALWANI and ACREE (1984), and ROELOFFS (1988).

The relationship between the time t required for pore pressure to cover the distance Δ is:

$$\Delta = \left(\frac{kt}{\gamma \Phi \beta} \right)^{1/2} \tag{2}$$

where k is permeability of the porous medium with dimensions of length squared, γ is fluid viscosity, Φ is the porosity of the fractured medium and β the effective fluid compressibility. The relationship (2) can be used to verify the trigger hypoth-

Table 1

Onset time for the anomalous increase and decrease in the flow rate

Earthquake	5/5/92	13/3/94	25/8/92
Water increase onset	19/4/92	4/1/94	30/3/92
Water decrease onset	30/4/92	29/1/94	2/5/92

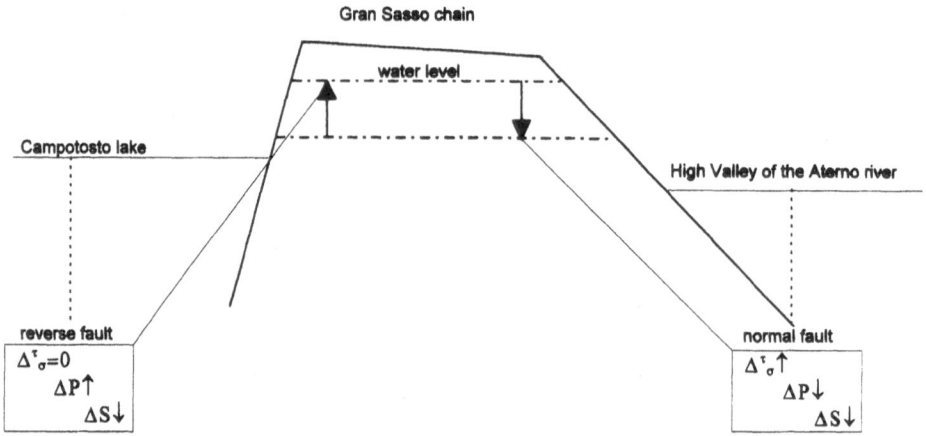

Figure 8
Schematic illustration of the destabilizing mechanism for the two earthquakes which occurred in the
Campotosto Lake zone (left side of Figure) and for the earthquake which occurred in the High Valley
of the Aterno River (right side of Figure).

esis for the two earthquakes of the Campotosto Lake area. If we assume that pore
pressure variations can transmit instantaneously inside the carbonatic structure, it
is possible to calculate distances just from the boundary zone (Fig. 5b). Assuming
$\gamma = 10^{-8}$ bar · s, $\Phi = 10^{-3}$ and $\beta = 3 \cdot 10^{-5}$ bar^{-1} and using the time values of the
water increase onset (Table 1), we obtain permeability values of $(0.5-1) \, 10^{-11}$ cm^2
for both cases; these values are in excellent agreement with those of the Laga
Mountains (see Section 2). Besides, in studying reservoirs, abrupt increases in water
level are considered the most destabilizing factor for reverse faults at an edge of the
reservoir (ROELOFFS, 1988) and the circumstance we have discussed in this paper
can be described by this phenomenology. An illustration of the quoted destabilizing
mechanism is shown on the left of the scheme reported in Figure 8.

We consider the earthquake in the High Valley of the Aterno River. The
hypocenter can be located at the margin of the Gran Sasso aquifer. Thus, the most
destabilizing factor in the relationship (1) is the stress σ and τ variation induced by
the water level changes of the aquifer (ROELOFFS, 1988). If changes of some 30–40
meters of the potentiometric surface of the Gran Sasso aquifer occur, values for the
destabilizing factor of some tenths of MPa are obtained and this is the same size
order calculated for the stress drops of the seismic sequence of the considered
earthquake. The fact that the earthquake took place at the end of the hydrologic
variation suggests that this seismic event is induced by the discharge process. As
regards a reservoir, normal faults beneath the reservoir are mostly destabilized by
means of a water level decrease (ROELOFFS, 1988) and the case we have just
presented seems to be an example of this phenomenology. An illustration of the
quoted destabilizing mechanism is shown on the right of the scheme reported in
Figure 8.

6. Conclusions

The evidence presented in this paper appears to indicate that the strongest earthquakes which occurred at the northwestern border area of the Gran Sasso basin during the last five years are induced ones. This induced effect has appeared clearly because of the particular meteorological condition that provoked abrupt and irregular variations in the aquifer. The possibility also exists that some minor seismic events can be numbered among the induced ones although it is extremely difficult to verify this hypothesis.

The hydrologic anomalies we have presented seem to be evidence of the cause of earthquakes and not of the stress readjustment processes. It seems important to verify this possibility when hydrologic anomalies are under study. These types of anomalies can then contribute to the improvement in the knowledge of the physics of earthquakes.

Acknowledgments

The authors are very grateful to the Ruzzo Waterwork Society management for the hydrologic data, and to the L'Aquila Tourist Board and National Meteorologic Service for snowfall and meteorologic data.

The authors thank also Drs. G. De Luca and G. Milana of the National Seismic Service for their willingness in providing seismic data, and Dr. R. Ranieri of ENEL for useful information concerning Campotosto Lake and local seismicity.

Particular appreciation is to our technical staff, P. Basili, F. Basti, and T. Coppola for their valuable cooperation during this research.

The authors extend gratitude to D. Simpson and P. Talwani for their constructive suggestions and comments in reviewing the manuscript.

REFERENCES

A.N.A.S.-CO.GE.FAR. (1980), Gran Sasso. Il Traforo autostradale, Edizioni Grafiche, Milano.

ANDERSON, R. E., and LANEY, R. L. (1975), *The Influence of Late Cenozoic Stratigraphy on Distribution of Impoundment-related Seismicity at Lake Mead, Nevada-Arizona*, J. Res. U.S. Geol. Surv. 3, 337–343.

BELL, M. L., and NUR, A. (1978), *Strength Changes due to Reservoir-induced Pore Pressure and Stresses and Application to Lake Oroville*, J. Geophys. Res. *83*, 4469–4483.

BELLA, F., BELLA, R., BIAGI, P. F., CAPUTO, M., DELLA MONICA, G., ERMINI, A., PLASTINO, W., and SGRIGNA, V. (1994), *Artificial and Natural Electromagnetic Signals revealed During two Years in the Amare Cave (Central Italy)*, Annali di Geofisica *XXXVII*, 1131–1136.

BELLA, F., BIAGI, P. F., CAPUTO, M., DELLA MONICA, G., ERMINI, A., PLASTINO, W., SGRIGNA, V., and ZILPIMIANI, D. (1995), *Electromagnetic and Seismoacoustic Signals Revealed in Karst Caves (Central Italy)*, Nuovo Cimento *18C*, 19–32.

BONGIOVANNI, G., CAPUANO, P., DE LUCA, G., MARSAN, P., and SCARPA, R. (1993), *Sismicità dell'Abruzzo Centrale mediante una rete sismica digitale*, Atti del XII Convegno del Gruppo Nazionale di Geofisica della Terra Solida, Roma 1993.

BONI, C., BONO, P., and CAPELLI, G. (1986), *Schema idrogeologico dell'Italia centrale*, Mem. Soc. Geol. It. *35*, 991–1012.

GHISETTI, F. (1987), *Mechanism of Thrust Faulting in the Gran Sasso Chain (Central Apennines, Italy)*, J. Struct. Geol. *9*, 955–966.

GHISETTI, F., and VEZZANI, L. (1991), *Thrust Belt Development in the Central Apennines (Italy): Northward Polarity of Thrusting and out-of-sequence Deformations in the Gran Sasso Chain*, Tectonics *10*, 904–919.

GOUGH, D. I., and GOUGH, W. I. (1976), *Time Dependence and Trigger Mechanisms for the Kariba (Rhodesia) Earthquakes*, Eng. Geol. *10*, 211–217.

ISTITUTO NAZIONALE DI GEOFISICA (1956–1995), Seismological Bulletins. Roma, Italy.

KEITH, C. M., SIMPSON, D. W., and SOBOLEVA, O. V. (1982), *Induced Seismicity and Style of Deformation at Nurek Reservoir*, Tadjik SSR, J. Geophys. Res. *87*, 4609–4624.

RICE, J. R., and CLEARY, M. P. (1976), *Some Basic Stress Diffusion Solutions for Fluid-saturated Elastic Porous Media with Compressible Constitutens*, Rev. Geophys. *14*, 227–241.

ROELOFFS, E. A. (1988), *Fault Stability Changes Induced beneath a Reservoir with Cyclic Variations in Water Level*, J. Geophys. Res. *93*, 2107–2124.

ROGERS, A. M., and LEE, W. H. K. (1976), *Seismic Study of Earthquakes in the Lake Mead, Nevada-Arizona Region*, Bull. Seismol. Soc. Am. *66*, 1657–1681.

SIMPSON, D. W., and NEGMATULLAEV, S. K. (1981), *Induced Seismicity at Nurek Reservoir, Tadjikistan, USSR*, Bull. Seismol. Soc. Am. *71*, 1561–1586.

SNOW, D. T. (1982), *Hydrogeology of Induced Seismicity and Tectonism: Case Histories of Kariba and Koyna*, Spec. Pap. Geol. Soc. Am., 189.

TALWANI, P., and ACREE, S. (1984), *Pore Pressure Diffusion and the Mechanism of Reservoir-induced Seismicity*, Pure appl. geophys. *122* (6), 947–965.

(Received July 30, 1997, revised September 9, 1997, accepted October 20, 1997)

Seismicity Due to Steam Stimulation and Oil Extraction

Pure appl. geophys. 153 (1998) 197–217
0033–4553/98/010197–21 $ 1.50 + 0.20/0

| Pure and Applied Geophysics

A Seismic Model of Casing Failure in Oil Fields

S. Talebi,[1] T. J. Boone[2] and S. Nechtschein[2]

Abstract—We develop a seismic model that characterises the sudden tensional failure of oil-well casings. The energy released by the rupture of a well casing is transformed into heat and seismic energy. The upper bound of the seismic efficiency of this process is estimated at about 3%. The static situation at the completion of a casing failure episode is modelled by calculating the static displacement field generated by two opposing forces separated by an arm. The azimuthal patterns of these displacements and the change in the strain and stress fields caused by the force couple are described. The dynamics of the failure episode are modelled as a dipole with a seismic moment equivalent to the product of the average drop in shear stress, the failure surface, and an arm. The radiated P and S waves have mean-square radiation pattern coefficients of $1/5$ for P waves and $2/15$ for S waves. The displacement field as a function of time during rupture and the spectral properties in the far field are derived. The most promising seismic parameters that can be used for distinguishing between casing failure events and other possible events are polarisation properties of S waves and S/P amplitude ratios. S-wave polarisation distinguishes between shear events and casing failure events. S/P amplitude ratios distinguish between tensile events and casing failure events.

Key words: Casing failure, oil sands, seismic modelling, steam stimulation.

Introduction

The most popular application of seismic monitoring techniques in oil fields aims at detecting the geometry of hydraulic fractures induced by fluid injections (e.g., Power *et al.*, 1976; Lacy, 1987; Talebi *et al.*, 1991). Other applications include monitoring hydraulic stimulation for secondary recovery of oil (e.g., Raleigh *et al.*, 1972; Phillips *et al.*, 1998), masive fluid withdrawals (e.g., Segall, 1989; McGarr, 1991; Doser *et al.*, 1992), steam injections and fireflood monitoring (Dusseault and Nyland, 1982; Gendzwill, 1992). Nicholson and Wesson (1992) have presented an extensive review of seismicity triggered by deep well activities.

[1] CANMET, 1079 Kelly Lake Rd., Sudbury, Ontario, Canada P3E 5P5. Fax: (705) 670-6556. E-mail: stalebi@nrcan.gc.ca
[2] Imperial Oil Resources Ltd., 3535 Research Rd. N.W., Calgary, Alberta, Canada T2L 2K8.

Hydrocarbon recovery in heavy oil reservoirs requires particular procedures in order to overcome the large viscosity of the bitumen and allow its flow towards producing wells. Cyclic Steam Stimulation (CSS) is one such procedure used at Imperial Oil Ltd.'s Cold Lake oil field in Alberta (e.g., BUCKLES, 1979; VIT-TORATOS *et al.*, 1990). The procedure, developed through an extensive research program, consists of injecting large volumes of steam under high pressures and temperatures, alternating with oil and water production for as many cycles as economic conditions permit (KRY, 1989). The high fluid pressures inside well casings used for this operation and the cyclic nature of thermal loading due to episodes of high and low fluid pressure and temperature can result in failure of borehole casings in pure tension. Early detection of such failures is critical from an operational point of view. Seismic monitoring techniques have the potential to identify such failures. To realise this potential, it is necessary to understand the seismic radiation due to this process. This paper presents a seismic model that has been successfully applied to *in situ* seismic data from the Cold Lake oil field (TALEBI *et al.*, 1998b).

Preliminary Considerations

Consider a section of a borehole casing at a depth D before failure. The normal stress σ_n applied to the outside of the casing by the rock mass after its installation and before the cement sets is equal to the pressure of the cement between the casing and the rock:

$$\sigma_n = \rho_C gD, \tag{1}$$

where ρ_C is the density of the cement and g is the acceleration due to gravity. This normal stress could approach the average horizontal stress levels within the rock mass some time after the casing installation. The value will also depend on the degree of shrinkage as the cement cures. The normal stress will fluctuate during steaming, shut-in and production intervals due to changing fluid pressures and temperatures inside the casing. Nonetheless, Equation (1) is used as an approximate measure of the normal stress on the casing. No frictional stresses are considered to be present between the casing and the rock mass prior to the failure episode.

After failure in tension of the casing along a circular section at depth D, the two ends start moving away from each other and frictional forces are mobilised. It will take a finite amount of time for the two sliding sections to mobilise enough frictional force to match the tensional force within the casing. A balance is achieved after the slippage of a sufficiently large section of the casing along a length L at each end. The Coulomb failure criteria can be expressed as

$$|\tau|_{\text{failure}} = C + \mu_i \sigma_n, \tag{2}$$

where τ is the frictional stress and μ_i represents the coefficient of friction. One assumes that there is no cohesion between the casing and the rock because of the repeated cycles of thermal loading (i.e., $C = 0$). The shear stress between the casing and the rock at failure initiation is

$$\tau_s = \mu_s \sigma_n, \tag{3}$$

where μ_s is the static coefficient of friction. The increase of frictional stresses from zero before the casing failure to the static shear stress estimated from Equation (3) has to be achieved before sliding can initiate at a point. It is generally accepted that after a period of time, the shear stress at each slipping point decreases to the dynamic shear stress

$$\tau_d = \mu_d \sigma_n. \tag{4}$$

The dynamic coefficient of friction μ_d has been observed to be about 90% to 95% of its static counterpart (JAEGAR and COOK, 1979). One assumes that at the end of the slippage the shear stresses at the two failed ends of the casing can be estimated from the above equation. With a P-wave velocity of 5 km/s for steel, the rupture of steel can be assumed to take place at a maximum speed of about 1 km/s (BROEK, 1986). Slippage between the rock and the casing is expected to have a speed in the order of the Rayleigh-wave velocity for the rock (about 0.5 km/s for shales). In a practical situation, the failure of the casing itself is expected to be about two orders of magnitude faster than the subsequent slippage episode against the wall (TALEBI, 1998), i.e., the casing breaks almost instantly. Figure 1 shows the situation at the end of the process when the coefficient of friction at the end of the casing has dropped to μ_d while it is equal to μ_s at the tip of the rupture front. The force of tension within the casing in areas away from the failure area remains unchanged before and after this process and can be estimated from

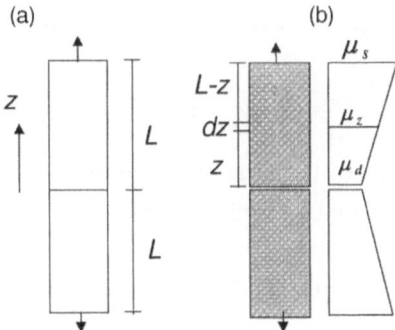

Figure 1

The conditions before and after tensional failure of a cylindrical casing: a) The circular section in the plane $z = 0$ is about to break. b) After the completion of failure, the forces of friction along the shaded area of the casing match the forces of tension within the casing. The coefficient of friction is μ_d for $z = 0$, μ_s for $z = L$ and $z = -L$.

$$F_c = \pi dBfr\sigma_y, \tag{5}$$

where d is the diameter and B is the thickness of the casing, fr is the fraction of the section that breaks suddenly, and σ_y is the yield stress of the steel. The total force of friction mobilised between the rock and the casing is

$$F_f = \tfrac{1}{2}\pi dL(\tau_d + \tau_s). \tag{6}$$

Adopting a factor to express the loss of friction (Fig. 1)

$$\gamma = \frac{\mu_s - \mu_d}{\mu_s} = 1 - \frac{\mu_d}{\mu_s}, \tag{7}$$

results in:

$$\mu_d = (1 - \gamma)\mu_s. \tag{8}$$

Equilibrium is achieved when the force of friction matches the tensional force within the casing. Equating (5) and (6) yields an estimate of the length L as follows

$$L = \frac{2Bfr\sigma_y}{\mu_s\sigma_n(2 - \gamma)}. \tag{9}$$

Note that this length is inversely proportional to the normal stress on the casing.

Energy Balance

The seismic energy released as a result of casing failure is a fraction of the strain energy within the two sliding sections of the casing (Fig. 1). The released energy W_r is consumed mainly via two processes: production of heat along the sliding surfaces, W_h, and radiation of seismic waves, W_s. The principle of conservation of energy requires that

$$W_r = W_h + W_s. \tag{10}$$

Because of the symmetry of the situation, consider only the top section. The shear stress between the casing and the wall as a function of z is approximated as follows

$$\tau(z) = \left[\mu_d + (\mu_s - \mu_d)\frac{z}{L}\right]\sigma_n. \tag{11}$$

The term in square brackets is simply the coefficient of friction at z. This equation can be rewritten using Equations (7) and (8)

$$\tau(z) = \left[1 - \gamma\left(1 - \frac{z}{L}\right)\right]\mu_s\sigma_n. \tag{12}$$

The force $dF(z)$ between the wall and a small cylindrical fraction of the casing dz, located at z, can be expressed as

$$dF(z) = \tau(z)\pi\, d\, dz. \tag{13}$$

Integration of this quantity over the section from 0 to z gives the total force applied by the rock to this section

$$F(z) = \pi d\mu_s \sigma_n \left[(1-\gamma)z + \frac{\gamma}{2L} z^2 \right]. \tag{14}$$

Stresses and strains in the casing as a function of z can then be calculated from linear elasticity

$$\sigma(z) = \frac{\mu_s \sigma_n}{B} \left[(1-\gamma)z + \frac{\gamma}{2L} z^2 \right], \tag{15}$$

$$\varepsilon(z) = \frac{\mu_s \sigma_n}{BE} \left[(1-\gamma)(L-z) + \frac{\gamma}{2L} (L^2 - z^2) \right], \tag{16}$$

where E is the Young's modulus. The displacement of each section of the casing is given by

$$\delta(z) = \int_z^L \varepsilon(z)\, dz = \frac{\sigma_n \mu_s}{6BE} \left[3(1-\gamma)(L^2 - z^2) + \frac{\gamma}{L} (L^3 - z^3) \right]. \tag{17}$$

For example, by replacing z with L, the displacement at the rupture front is zero, as expected. The displacement at the two ends where failure initiated is obtained by putting $z = 0$

$$\delta(o) = \frac{\mu_s \sigma_n L^2}{6BE} (3 - 2\gamma). \tag{18}$$

Following the methodology of SALAMON (1993), the energy components are defined as a function of frictional stresses before and after failure

$$W_r = \frac{1}{2} \iint_\Sigma [\tau_s + \tau(z)]\, \delta(z)\, d\Sigma, \tag{19}$$

$$W_h = \iint_\Sigma [\tau(z)]\, \delta(z)\, d\Sigma, \tag{20}$$

$$W_s = \frac{1}{2} \iint_\Sigma [\tau_s - \tau(z)]\, \delta(z)\, d\Sigma. \tag{21}$$

The integration is performed over the sliding surface Σ where shear movement has taken place. The results are as follows

$$W_r = \frac{\pi d\sigma_n^2 \mu_s^2 L^3}{120BE} (80 - 75\gamma + 16\gamma^2), \tag{22}$$

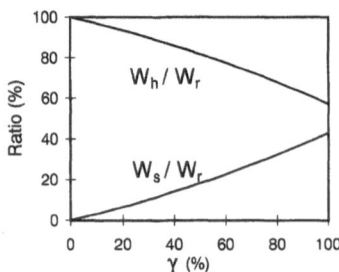

Figure 2
The ratio of the energy consumed in heat W_h and seismic radiation W_s over the total released energy W_r
as a function of the coefficient γ.

$$W_h = \frac{\pi d\sigma_n^2 \mu_s^2 L^3}{30BE} (20 - 25\gamma + 8\gamma^2),\qquad(23)$$

$$W_s = \frac{\pi d\sigma_n^2 \mu_s^2 L^3}{120BE} \gamma(25 - 16\gamma).\qquad(24)$$

The seismic energy in this model is a consequence of the decrease in the coefficient of friction. Figure 2 shows the proportions of these energy components as a function of γ. Note that for a maximum value of $\gamma = 0.1$, the ratio of seismic energy over the total released energy, commonly called seismic efficiency, has an upper bound of about 3% for this model.

Static Problem

In this section, one determines the static displacement field at any point within an isotropic homogeneous solid due to two opposing forces acting on the solid. The force couple represented in this situation is similar to the forces applied to the rock mass after a casing failure episode.

Displacement Field

First, consider the solution to the problem of a single force F defined as

$$F = \lim_{\delta V \to 0} \rho f \, \delta V,\qquad(25)$$

where f is force per unit mass, ρf is the body force per unit volume and δV is a small element of the body. This presentation and the introduction of the three-dimensional delta function allows for the basic elastic equations of equilibrium to be resolved (LAY and WALLACE, 1995)

$$F + (\lambda + 2\mu)\nabla(\nabla \cdot U) - \mu\nabla \times \nabla \times U = 0.\qquad(26)$$

The result is the Somigliana tensor

$$U_i^j = \frac{1}{8\pi\mu} (\delta_{ij} r_{,kk} - \Gamma r_{,ij}),$$ (27)

where

$$\Gamma = \frac{\lambda + \mu}{\lambda + 2\mu}.$$ (28)

Equation (27) gives the i-th component of displacement for a unit force ($F = 1$) in the j-th direction. For a Poisson solid, $\lambda = \mu$; and, therefore, $\Gamma = 2/3$. In our case, the two forces are along the z direction with the arm also along the same direction. If one takes the product $M = F. \Delta z$ as a finite quantity defined as a moment, and let $\Delta z \to 0$ and $F \to \infty$, the displacement field due to this force couple is expressed as follows

$$U_1 = -\frac{\partial U_1^1}{\partial z},$$ (29)

$$U_2 = -\frac{\partial U_2^1}{\partial z},$$ (30)

$$U_3 = -\frac{\partial U_3^1}{\partial z}.$$ (31)

In other words, the derivative of equation (27) relative to z yields the displacement field for the force couple at any given point in the medium. The results of this operation in a Cartesian coordinated system (xyz) are as follows

$$U_x = -\frac{M}{8\pi\mu} \left[\Gamma \left(\frac{x}{r^3} - \frac{3xz^2}{r^5} \right) \right],$$ (32)

$$U_y = -\frac{M}{8\pi\mu} \left[\Gamma \left(\frac{y}{r^3} - \frac{3yz^2}{r^5} \right) \right],$$ (33)

$$U_z = -\frac{M}{8\pi\mu} \left[-\frac{2z}{r^3} - 3\Gamma \left(-\frac{z}{r^3} + \frac{z^3}{r^5} \right) \right].$$ (34)

It is more convenient to express these results in cylindrical coordinates (Fig. 3) as a function of ($r\theta z$) via the use of the following transformation matrix

$$\begin{vmatrix} U_r \\ U_\theta \\ U_z \end{vmatrix} = \begin{vmatrix} \cos\theta & \sin\theta & 0 \\ -\sin\theta & \cos\theta & 0 \\ 0 & 0 & 1 \end{vmatrix} \begin{vmatrix} U_x \\ U_y \\ U_z \end{vmatrix}.$$ (35)

Hence:

$$U_r = -\frac{M}{8\pi\mu r^2}\left[\Gamma\left(1 - \frac{3z^2}{r^2}\right)\right], \tag{36}$$

$$U_\theta = 0, \tag{37}$$

$$U_z = -\frac{M}{8\pi\mu r^2}\frac{z}{r}\left[3\Gamma\left(1 - \frac{z^2}{r^2}\right) - 2\right]. \tag{38}$$

Two aspects of these results could have been expected from the start because of the symmetry of the problem. One is the fact that $U_\theta = 0$ everywhere, and the other is that the vertical displacement $U_z = 0$ for the plane $z = 0$.

Focusing on oil applications where rocks have typically much higher Poisson's ratios, values of 0.34 and 0.44 are considered here as reasonable estimates of Poisson's ratios for sandstone and shale. This results in a ratio of P/S wave velocities of 2 and 3. Taking the latter value which is more representative of our application (TALEBI *et al.*, 1998a) and rewriting the equation (28)

$$\Gamma = 1 - \frac{\mu/\rho}{(\lambda + 2\mu)/\rho} = 1 - \left(\frac{V_S}{V_P}\right)^2 = \frac{8}{9}. \tag{39}$$

Then Equations (36) and (38) can be rewritten:

$$U_r = -\frac{M}{9\pi\mu r^2}\left(1 - \frac{3z^2}{r^2}\right), \tag{40}$$

$$U_z = -\frac{M}{12\pi\mu r^2}\left[\frac{z}{r}\left(1 - \frac{4z^2}{r^2}\right)\right]. \tag{41}$$

An even more convenient manner of looking at the directional properties of the displacement field is to use spherical coordinates (Fig. 3) using the Jacobian coordinate transformation

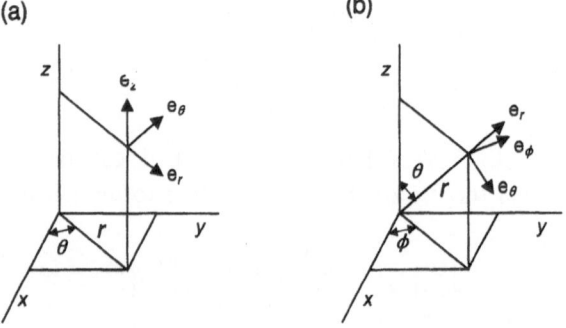

Figure 3
a) Cylindrical (r θ z) and b) spherical (r θ ϕ) systems of coordinates.

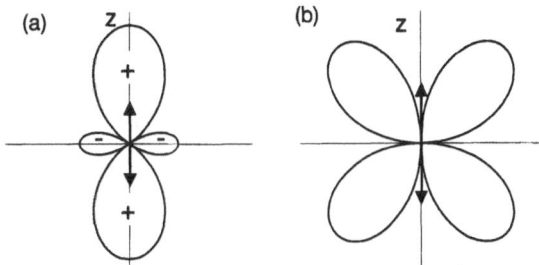

Figure 4
Azimuthal patterns of static displacements in the vertical plane for a dipole: a) Radial displacement; b) tangential displacement.

$$\begin{vmatrix} U_r \\ U_\theta \\ U_\phi \end{vmatrix} = \begin{vmatrix} \sin\theta\cos\phi & \sin\theta\sin\phi & \cos\theta \\ \cos\theta\cos\phi & \cos\theta\sin\phi & -\sin\theta \\ -\sin\phi & \cos\phi & 0 \end{vmatrix} \begin{vmatrix} U_x \\ U_y \\ U_z \end{vmatrix}. \tag{42}$$

In the vertical plane xz, $\phi = 0$ and $y = 0$, then we find

$$U_r = U_x \sin\theta + U_z \cos\theta, \tag{43}$$

$$U_\theta = U_x \cos\theta - U_z \sin\theta, \tag{44}$$

$$U_\phi = 0. \tag{45}$$

resulting in the following equations in spherical coordinates for the displacements in the vertical plane (also replacing Γ from Equation (28))

$$U_r = \frac{M}{72\pi\mu r^2}(5 + 13\cos 2\theta), \tag{46}$$

$$U_\theta = -\frac{M}{72\pi\mu r^2}\sin 2\theta. \tag{47}$$

Figure 4 shows the azimuthal patterns of these displacements in the vertical plane.

Strain and Stress Fields

The displacement field obtained in the previous section can be applied to calculate the strain and stress fields caused by casing failure using linear elasticity. The cylindrical representation is utilized since $U_\theta = 0$ and equations are easier to handle. Moreover, only four derivatives of the displacement field are non-zero and need to be considered

$$\frac{\partial U_r}{\partial r} = -\frac{M\Gamma}{8\pi\mu}\left(-\frac{2}{r^3} + \frac{12z^2}{r^5}\right), \tag{48}$$

$$\frac{\partial U_r}{\partial z} = -\frac{M\Gamma}{8\pi\mu}\left(-\frac{6z}{r^4}\right), \tag{49}$$

$$\frac{\partial U_z}{\partial r} = -\frac{M}{8\pi\mu}\left(-\frac{9\Gamma z}{r^4} + \frac{15\Gamma z^3}{r^6} + \frac{6z}{r^4}\right), \tag{50}$$

$$\frac{\partial U_z}{\partial z} = -\frac{M}{8\pi\mu}\left(\frac{3\Gamma}{r^3} - \frac{9\Gamma z^2}{r^5} - \frac{2}{r^3}\right). \tag{51}$$

As a result of this, all three components of the strain field related to θ are zero and only the following three components remain

$$\varepsilon_{rr} = -\frac{M\Gamma}{8\pi\mu r^3}\left(-2 + 12\frac{z^2}{r^2}\right), \tag{52}$$

$$\varepsilon_{rz} = -\frac{M}{8\pi\mu r^3}\frac{z}{r}\left[\frac{15\Gamma}{2}\left(\frac{z^2}{r^2} - 1\right) + 3\right], \tag{53}$$

$$\varepsilon_{zz} = -\frac{M}{8\pi\mu r^3}\left[(3\Gamma - 2) - 9\Gamma\frac{z^2}{r^2}\right]. \tag{54}$$

All components of the stress field can now be calculated from Hooke's law

$$\sigma_{ij} = \lambda\delta_{ij}(\text{tr } \varepsilon) + \mu\varepsilon_{ij}. \tag{55}$$

For example, consider the horizontal plane $z = 0$

$$\varepsilon_{rr} = \frac{M}{8\pi\mu r^3}(2\Gamma), \tag{56}$$

$$\varepsilon_{rz} = 0, \tag{57}$$

$$\varepsilon_{zz} = \frac{M}{8\pi\mu r^3}(2 - 3\Gamma). \tag{58}$$

The trace of the strain tensor then is expressed as

$$\text{tr } \varepsilon = \varepsilon_{rr} + \varepsilon_{zz} = \frac{M}{8\pi\mu r^3}(2 - \Gamma). \tag{59}$$

Using Equation (55), one can obtain the change of stress in the plane $z = 0$

$$\sigma_{rr} = \frac{M}{8\pi\mu r^3}[2\lambda + (4\mu - \lambda)\Gamma], \tag{60}$$

$$\sigma_{zz} = \frac{M}{8\pi\mu r^3}[2(\lambda + 2\mu) - \Gamma(\lambda + 6\mu)]. \tag{61}$$

Dynamic Problem

The dynamic problem is clearly more difficult since the variation of forces with time must be considered. There is an analogy in the approach with the static case in that the classical Stoke's solution for a point force is used as a basis (e.g., AKI and RICHARDS, 1980; Equations 4.23). This solution is then used, as the Somigliana tensor was used in the static case, to derive the dynamic solution. The complete solution for any seismic source is given by AKI and RICHARDS (1980; Equations 4.29). The only relevant components of seismic radiation are the far field terms since receivers are never located within a few wavelength from the source. Therefore, the so-called near field and intermediate field terms become negligible; i.e.,

$$U_P = \frac{\gamma_n \gamma_p \gamma_q}{4\pi\rho\alpha^3} \frac{\dot{M}_{pq}(t - r/\alpha)}{r}, \tag{62}$$

$$U_S = \frac{(\delta_{np} - \gamma_n \gamma_p)\gamma_q}{4\pi\rho\beta^3} \frac{\dot{M}_{pq}(t - r/\beta)}{r}, \tag{63}$$

where α and β represent P- and S-wave velocities, M_{pq} defines the component pq of the moment tensor of the source and the dot sign represents the derivative with respect to time. Two important aspects of this source that must be considered are seismic moment and radiation pattern.

Seismic Moment

The seismic moment is derived for the source represented by Figure 5. Assuming that failure starts at $z = 0$ and proceeds at a constant velocity V_R until it stops at $z = L$ and $-L$, one needs to consider body-force equivalents for a discontinuity of traction along a surface. Adopting the methodology of AKI and RICHARDS (1980, Equation 3.4), for the present case, the surface considered is a cylinder, all displacements within the solid are continuous, and body forces are in the z direction:

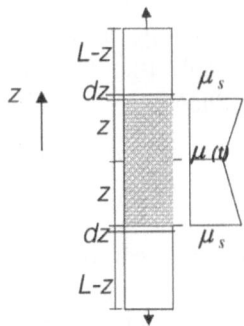

Figure 5

The situation at time t after the initiation of sliding between the casing and the wall. At each end of the casing, a section with a length z is sliding (shaded area). The rupture front progresses upwards and downwards until the remaining parts $(L - z)$ are broken and equilibrium is achieved.

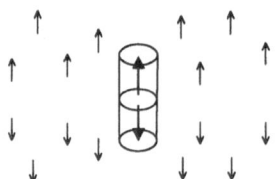

Figure 6
Body force equivalents for the dipole source.

$$f^{\prime\eta}(\boldsymbol{\eta}, \tau) = -\iint_{\Sigma}[T(U(\boldsymbol{\xi}, \tau), \nu)]\delta(\boldsymbol{\eta} - \boldsymbol{\xi})\, d\Sigma(\boldsymbol{\xi}), \qquad (64)$$

where square brackets represent the difference of a parameter on opposing sides of the surface Σ with the traction discontinuity, ν is a vector normal to Σ, $\boldsymbol{\eta}$ is a vector pointing to a general position within the medium, $[T(U(\boldsymbol{\xi}, \tau), \nu)]$ is a vector corresponding to traction discontinuity at a given time τ and a given location $\boldsymbol{\xi}$ on the surface Σ, and δ is the three-dimensional delta function. This represents a system of force couples within the body with both forces and arms along the vertical axis (Fig. 6). The total sesimic moment is represented by the integration of these elementary seismic moments over the entire volume within the rock mass (AKI and RICHARDS, 1980). The results are equivalent to the following surface integral

$$M_{zz}(\tau) = \iint_{\Sigma}[T_z(\boldsymbol{\xi}, \tau)\, \delta(z)\, d\Sigma. \qquad (65)$$

Our seismic moment tensor has only one component M_{zz} since both body forces and arms are parallel to the vertical axis. Being interested in the total seismic moment at the end of rupture, consider Figure 5 to calculate the seismic moment when rupture is completed. The two rupture fronts initiate at $z = 0$ and proceed in opposite directions with a constant velocity V_R until they reach the coordinates L and $-L$. Assuming then that the surface element dz at location z starts undergoing dynamic movement at time 0 and stops moving at time $(L - z)/V_R$, the extent of the rupture front at time t is simply $V_R t$. The frictional stress applied to the rock at time t can be expressed as follows

$$\tau(t) = \mu_s\sigma_n\left(1 - \gamma\frac{V_R}{L}t\right). \qquad (66)$$

This linear dependence of frictional stress with time is in agreement with the model of PALMER and RICE (1973). Since seismic radiation results directly from the time derivative of elementary seismic moments, one takes the derivative of this quantity with respect to time, i.e.,

$$\dot{\tau}(t) = \mu_s\sigma_n\left(-\gamma\frac{V_R}{L}\right). \qquad (67)$$

In other words, stresses applied to the rock during failure are dropping at a constant rate, independent of time. Multiplying this quantity by the surface of this element and the arm yields the time derivative of the elementary seismic moment

$$\dot{m}(t) = 2\pi d d z z \mu_s \sigma_n \left(-\gamma \frac{V_R}{L} \right). \tag{68}$$

The total contribution of this element can be obtained if one integrates the above quantity with respect to time over its rupture time and take the absolute value. Note that this quantity is independent of time during the failure of the element since z remains constant during rupture. The result is

$$m_0 = 2\pi d d z z \gamma \mu_s \sigma_n \left(1 - \frac{Z}{L} \right). \tag{69}$$

The total seismic moment is obtained by integrating this quantity over z from zero to L, i.e.,

$$M_0 = \tfrac{1}{3} \pi d \gamma \mu_s \sigma_n L^2. \tag{70}$$

Note this is simply equivalent to the product of the average drop in shear stress $\gamma \tau_s / 2$, the failure surface $\pi d L$, and the length of the arm $2L/3$.

Radiation Pattern

It is more convenient to express P- and S-wave displacement fields (Equations (62) and (63)) in spherical coordinates. Following the methodology of GIBOWICZ and KIJKO (1994), the far-field displacement field is decomposed into P, SV and SH components:

$$U^P = \frac{1}{4\pi\rho\alpha^3} \frac{1}{r} R^P \mathbf{R}, \tag{71}$$

$$U^{SV} = \frac{1}{4\pi\rho\beta^3} \frac{1}{r} R^{SV} \boldsymbol{\theta}, \tag{72}$$

$$U^{SH} = \frac{1}{4\pi\rho\beta^3} \frac{1}{r} R^{SH} \boldsymbol{\phi}. \tag{73}$$

As mentioned in the previous section, the seismic moment in this case has only one component M_{zz}. Following the methodology of the above authors, the source radiation terms can be derived as follows

$$R^P = \cos^2 \theta \dot{M}_{zz}(t - r/\alpha), \tag{74}$$

$$R^{SV} = -\tfrac{1}{2} \sin 2\theta \dot{M}_{zz}(t - r/\beta), \tag{75}$$

$$R^{SH} = 0. \tag{76}$$

One important consequence of these results is that the radiated S wave in the present model consists of an SV component only, as there is symmetry around the vertical axis. Figure 7 shows radiation patterns of P and SV waves for the present model. In order to use Equations (74) to (76) for the cases in which the direction of the dipole is not known, one needs to calculate the mean square value of the above quantities. Averaging over the focal sphere results in

$$\langle (\cos^2 \theta)^2 \rangle = \frac{1}{2} \int_0^\pi (\cos^4 \theta) \sin \theta \, d\theta = \frac{1}{5}, \tag{77}$$

$$\left\langle \left(-\frac{1}{2} \sin 2\theta \right)^2 \right\rangle = \frac{1}{2} \int_0^\pi \left(-\frac{1}{2} \sin 2\theta \right)^2 \sin \theta \, d\theta = \frac{2}{15}. \tag{78}$$

The consequences of these results will be discussed later when comparing the properties of this model to those of shear and tensile failure within the rock mass.

Displacement Field

In this section, the methodology of HASKELL (1964), as described by LAY and WALLACE (1995), is followed. Considering Figure 5 again and cutting the cylinders of length L on each side into N equal pieces, one can calculate the displacement field radiated by each segment and total them to obtain the displacement field as a function of time. The main difficulty of this procedure from a practical point of view is the fact that each element starts its contribution at a given time, when the rupture front reaches it, but all the elements terminate their contributions at almost the same time when energy balance is reached. Consider the P-wave displacement field in cylindrical coordinates as follows

$$U_r(r, t) = \frac{1}{4\pi\rho\alpha^3} \frac{R^P}{r} \dot{M} \left(t - \frac{r}{\alpha} \right). \tag{79}$$

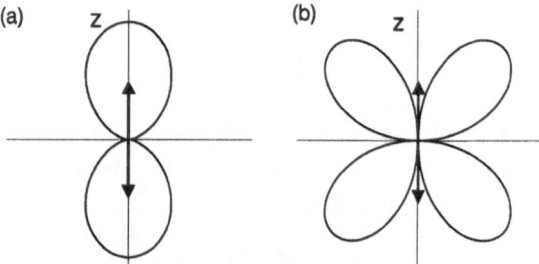

Figure 7
Radiation pattern of P waves (a) and S waves (b) for the present model. The arrows show the dipole which represents the casing failure source.

It will take a finite amount of time for each point to reduce its frictional stresses from the static level to a lower value according to Equation (66). The frictional stress has to reach the static value before rupture initiates and, after some time, the coefficient of friction decreases to its dynamic value. The time derivative of frictional stress expressed by the Equation (67) will map directly into seismic moment. Considering an element i at z_i, the elementary seismic moment was expressed by Equation (69). Since the time origins of these elements are different, one must consider this quantity within the new time scale as a function of time; i.e., let it intervene only during the time period when this element is radiating seismic energy with $\dot{\tau} = 0$ for $t - \Delta t_i \leq 0$. The summation of this contribution, considering the time lag between the elements, Δt_i, is as follows

$$U_r(r, t) = \frac{2\pi d}{4\pi\rho\alpha^3} \sum_{i=1}^{N} \frac{R_i^P z_i \dot{\tau}(t - \Delta t_i)}{r_i} \, dz. \tag{80}$$

Assume that the time lag Δt_i is equivalent to the extent of the rupture front along the vertical axis divided by the rupture velocity V_R. This would be the case for a sensor located remotely but at about the same depth as the failure area. In such a case, r_i and R_i^P are approximately constant and can be taken outside the summation

$$U_r(r, t) = \frac{2\pi d R^P}{4\pi\rho\alpha^3 r} \sum_{i=1}^{N} z_i \dot{\tau}(t - z_i/V_R) \, dz. \tag{81}$$

This equation can be rewritten using the shift property of the delta function

$$\dot{\tau}(t - z/V_R) = \dot{\tau}(t) * \delta(t - z/V_R), \tag{82}$$

where * denotes convolution and $\dot{\tau}(t)$ is the same everywhere along the shearing surface. Putting (82) into (81) and taking the limit of the sum as $dz \to 0$, one obtains the following integral equation

$$U_r(r, t) = \frac{2\pi d R^P}{4\pi\rho\alpha^3 r} \int_0^{z_f} z\dot{\tau}(t) * \delta(t - z/V_R) \, dz, \tag{83}$$

where z_f is the length of ruptured surface at time t. Since $\dot{\tau}(t)$ is independent of z, we take it outside the integral, i.e.,

$$U_r(r, t) = \frac{2\pi d R^P}{4\pi\rho\alpha^3 r} \dot{\tau}(t) * \int_0^{z_f} z\delta(t - z/V_R) \, dz. \tag{84}$$

One must make a change in variable in order to calculate the above integral:

$$u = t - z/V_R, \tag{85}$$

which results in the following

$$z = (t - u)V_R, \tag{86}$$

$$dz = -V_R \, du, \tag{87}$$

$$\int_0^{z_f} z\delta(t - z/V_R)\, dz = z_f V_R \int_{t - z_f/V_R}^t \delta(u)\, du. \tag{88}$$

The integral on the right side is a heavyside function; i.e., $H(t) = 0$ before time t and $H(t) = 1$ after time t. The displacement field is then

$$U_r(r, t) = \frac{2\pi d R^P}{4\pi\rho\alpha^3 r}\, \dot{\tau}(t) \ast z_f V_R H(u)\big|_{t - z/V_R}^t \tag{89}$$

$$= \frac{2\pi d R^P}{4\pi\rho\alpha^3 r}\, \dot{\tau}(t) \ast z_f V_R [H(t) - H(t - z_f/V_R)]. \tag{90}$$

The quantity in square brackets is a boxcar of duration equal to the rupture time z_f/V_R. Therefore, the shape of the far-field displacement according to this model is similar to the one derived by HASKEL (1964) and is defined by the convolution of two boxcars, one representing the displacement history of a single particle and the second describing the effects of a finite length, with durations τ_c and τ_r, respectively. In the present model τ_c and τ_r are equal. Integrating the far-field term over the entire time interval yields

$$\int_{-\infty}^{+\infty} U_r(r, t)\, dt = \int_{-\infty}^{+\infty} \frac{2\pi d R^P}{4\pi\rho\alpha^3 r}\, \dot{\tau}(t) \ast z_f V_R [B(t, z_f/V_R)]\, dt, \tag{91}$$

where $B(t, z_f/V_R)$ is the boxcar of duration equal to the duration of rupture; i.e., the term in brackets in Equation (90). Indeed, for the ends of the casing, z_f is equal to L at the completion of rupture. Multiplying both sides by the correction terms for geometrical spreading, radiation pattern and source constants results in

$$\frac{4\pi\rho\alpha^3 r}{R^P} \int_{-\infty}^{+\infty} U_r(r, t)\, dt = \int_{-\infty}^{+\infty} [2\pi d\dot{\tau}(t)] \ast [z_f V_R B(t, z_f/V_R)]\, dt. \tag{92}$$

The right-hand side of this equation corresponds to the total seismic moment at the end of rupture. The left-hand side is the area under the displacement pulse corrected for spreading, radiation pattern and source constants. This equation provides a procedure for the determination of seismic moment from far-field displacement signals in the time domain (see TALEBI *et al.*, 1998b).

Source Spectrum

The source time function from the previous equation can be rewritten as follows

$$U(t) = \Omega_0 [B(t, \tau_r) \ast B(t, \tau_c)]; \tag{93}$$

i.e., convolution of the two boxcars, as mentioned earlier. The Fourier transform of convolution of two boxcars in the time domain is simply equal to the multiplication of the Fourier transforms of the two functions in the frequency domain. Therefore, the Fourier transform of the above function can be written as

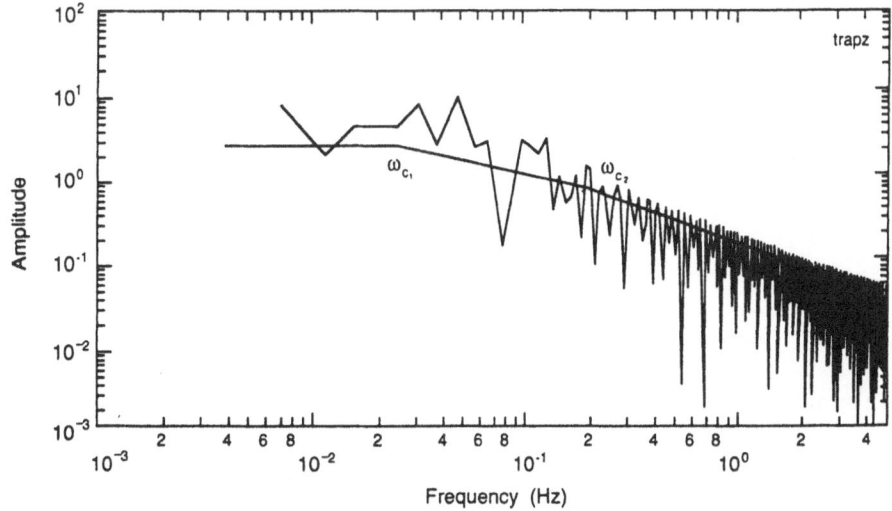

Figure 8
Amplitude spectrum of two boxcars manifests a flat trend at low frequencies and a descending trend with a slope of −2 at high frequencies (after LAY and WALLACE, 1995).

$$U(\omega) = \Omega_0 \left| \frac{\sin(\omega\tau_r/2)}{\omega\tau_r/2} \right| \left| \frac{\sin(\omega\tau_c/2)}{\omega\tau_c/2} \right|. \tag{94}$$

This function is clearly approaching an amplitude of Ω_0 (the amplitude of the function at zero frequency) at small frequencies; i.e., the spectrum has a plateau at small frequencies and then decays in proportion to ω^{-2} at frequencies larger than those corresponding to rupture duration (Fig. 8). Therefore, if the plateau level is defined from a classical Fast Fourier Transform, the seismic moment can be calculated in a similar manner as for shear models (e.g., BRUNE, 1970; MADARIAGA, 1976) or tensile models (e.g., SATO, 1978).

Discussion

The purpose of the present model is to allow the detection of casing failures in oil fields from their seismic signals. Other types of events originating from shear or tensile failure of rock must be distinguished from casing failures. The first piece of information to be considered is the source location of an event. Although this topic was not discussed so far, it is clear that only events within a critical distance of an oil well should be considered. Magnitude is another parameter that can be considered, but it would be of limited use because of the wide range of possible values that it can take for different types of events. However, this parameter should allow the exclusion of events considerably stronger than that which can be expected for a casing failure.

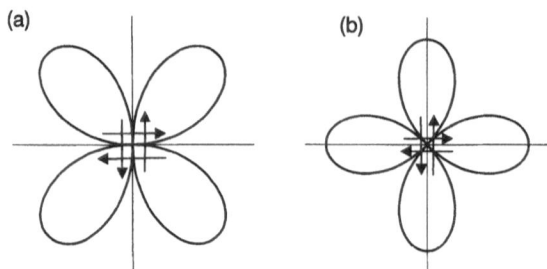

Figure 9
Radiation pattern of P waves (a) and S waves (b) for a shear event.

The radiation pattern of P and S waves has a large potential for solving the problem. The radiation pattern of body waves for shear and tensile events can be derived following the methodology in the previous section. For a shear event (i.e., double-couple event)

$$R^P = \sin 2\theta \cos \phi, \tag{95}$$

$$R^{SV} = \cos 2\theta \cos \phi, \tag{96}$$

$$R^{SH} = -\cos \theta \sin \phi, \tag{97}$$

and for a tensile event

$$R^P = (\alpha/\beta)^2 - 2 \sin^2 \theta, \tag{98}$$

$$R^{SV} = -\sin 2\theta, \tag{99}$$

$$R^{SH} = 0. \tag{100}$$

Figures 9 and 10 show these patterns assuming a value of 3 for the ratio of P/S-wave velocity. In order to use the differences between these patterns, the ideal solution would be to determine them from recorded signals. Note, for example, that the P-wave radiation pattern has two nodes for a shear event, one node for casing

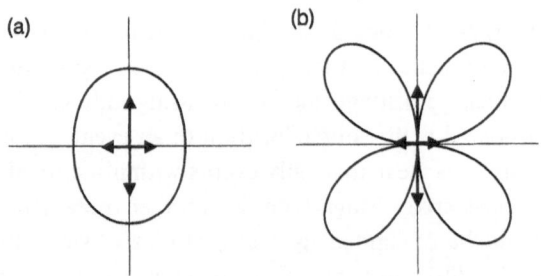

Figure 10
Radiation pattern of P waves (a) and S waves (b) for a tensile event.

failure and no nodes for tensile failure. Two parameters related to these radiation patterns would be of great potential: polarisation of S waves and S/P amplitude ratio.

The polarisation of S waves allows the discrimination of shear events from the two other types of events. Indeed, close examination of previous results indicates that the SH component is generated only in the case of shear failure. Obviously, at large distances, some SH component can be expected for all types of events because of the conversion of SV into SH at horizontal boundaries. This should not be a major problem with local arrays covering an oil pad where distances involved are in the range of 100–200 meters. Furthermore such effects would be limited when the source and the sensor are within the same layer. The presence of a strong SH component would then clearly indicate a shear event.

The average S/P amplitude or energy ratio is another strong parameter to be considered. Indeed, the total energy of each wave type is proportional to the average mean square value of the radiation factors. Repeating the procedure used in deriving such parameters for the case of casing failure in the case of shear events results in (AKI and RICHARDS, 1980)

$$\langle (R^P)^2 \rangle = \tfrac{4}{15}; \quad \langle (R^S)^2 \rangle = \tfrac{2}{3}, \tag{101}$$

and, for tensile events assuming a P/S velocity ratio of 3

$$\langle (R^P)^2 \rangle = \tfrac{887}{15}; \quad \langle (R^S)^2 \rangle = \tfrac{8}{15}. \tag{102}$$

Note that for a Poisson's solid, the only difference would be in the average mean square value for the P-wave radiation pattern for tensile events which is equal to 47/15 (e.g., WALTER and BRUNE, 1993). If one considers the ratio between the average mean square values for P and S waves for the three mechanisms involved, the result will be a ratio of 2/3 for shear failure, 3/2 for casing failure but 887/8 for tensile failure. In other words, tensile events generate considerably lower S-wave energies than the other two mechanisms. This should provide a reliable parameter which allows for the distinction of tensile events from casing failure events.

Conclusions

—The length of the sliding section at each end of a ruptured casing strongly depends on the shear stresses between the casing and the surrounding rock mass.

—Most of the energy released by the casing break is transformed into heat while seismic energy is a small fraction of the released energy. For a maximum of 10% decrease from the static to the dynamic coefficient of friction, the upper bound of seismic efficiency is estimated at about 3%.

—The seismic moment at the end of the rupture for the present model can be expressed, simply, as the product of the average stress drop, the failure surface, and the arm.

—The mean square values of radiation pattern coefficients from this model are 1/5 for *P* waves and 2/15 for *S* waves. Spectral characteristics of this seismic model are similar to those of the commonly used seismic models for shear and tensile failures: a constant trend at low frequencies and a descending trend at high frequencies with a −2 slope.

—Two potentially strong parameters that can be used for casing failure detection are polarisation of *S* waves and average *S/P* energy or amplitude ratio. The first parameter can allow the distinction between shear events and a casing failure event. The second parameter can be used to distinguish between tensile events and a casing failure event.

Acknowledgements

The authors would like to thank Imperial Oil Limited, Resources Division, for their financial support of this project and permission to publish the data. Reviews by Rick Kry and an anonymous reviewer helped to enhance the quality of this paper.

REFERENCES

AKI, K., and RICHARDS, P. G., *Quantitative Seismology* (W. H. Freeman, San Francisco 1980).
BROEK, D., *Elementary Engineering Fracture Mechanics* (Martinus Nijhoff, Dordrecht, The Netherlands 1986).
BRUNE, J. N. (1970), *Tectonic Stress and the Spectra of Seismic Shear Waves from Earthquakes*, J. Geophys. Res. *75*, 4997–5009.
BUCKLES, R. S. (1979), *Steam Stimulation Heavy Oil Recovery at Cold Lake, Alberta*, Proc. Calif. Meeting of the SPE of AIME, Ventura, Calif., April 18–20. SPE 7994, 1–12.
DOSER, D. I., BAKER, M. R., LUO, M., MARROQUIN, P., BALLESTEROS, L., KINGWELL, J., DIAZ, H. L., and KAIP, G. (1992), *The not so Simple Relationship between Seismicity and Oil Production in the Permian Basin, West Texas*, Pure appl. geophys. *139*, 481–506.
DUSSEAULT, M. B., and NYLAND, E., *Fireflood microseismic monitoring—rock mechanics implications*. In Proc. 23rd Symp. Soc. Min. Eng. (Am. Inst. Min. Metall. Petr. Eng., NY 1982).
GENDZWILL, D. J. (1992), *Passive Monitoring of Hydraulic Fracture and Steam Injection in Heavy Oil Sands near Fort McMurray, Alberta, Can.*, J. Expl. Geophys. *28*, 117–127.
GIBOWICZ, S. J., and KIJKO, A., *An Introduction to Mining Seismology* (Academic Press, San Diego 1994).
HASKELL, N. A. (1964), *Total Energy and Energy Spectra Density of Elastic Waves from Propagating Faults*, Bull. Seismol. Soc. Am. *54*, 1811–1841.
JAEGAR, J. C., and COOK, N. G. W., *Fundamentals of Rock Mechanics* (Chapman and Hall, London 1979).
KRY, P. R. (1989), *Field observations of steam distribution during injection to the Cold Lake reservoir*. In Proc. Symp. *Rock at Great Depth* (Pau, France) pp. 853–861.
LACY, L. L. (1987), *Comparison of Hydraulic-fracture Orientation Techniques*, SPE formation Evaluation, March 1987, pp. 66–76.
LAY, T., and WALLACE, T. C., *Modern Global Seismology* (Academic Press, San Diego 1995).
MADARIAGA, R. (1976), *Dynamics of an Expanding Circular Fault*, Bull. Seismol. Soc. Am. *66*, 639–666.

McGARR, A. (1991), *On a Possible Connection between Three Major Earthquakes in California and Oil Production*, Bull. Seismol. Soc. Am. *81*, 948–970.

NICHOLSON, C., and WESSON, R. L. (1992), *Triggered Earthquakes and Deep Well Activities*, Pure appl. geophys. *139*, 561–578.

PALMER, A. C., and RICE, J. R. (1973), *The Growth of Slip Surfaces in the Progressive Failure of Overconsolidated Clay Slopes*. In Proc. Roy. Soc. London *A332*, 527–548.

PHILLIPS, W. S., FAIRBANKS, T. D., RUTLEDGE, J. T., and ANDERSON, D. W. (1998), *Induced Microearthquake Patterns and Oil-producing Fracture Systems in the Austin Chalk*, Tectonophysics *289*, 153–169.

POWER, D. V., SCHUSTER, C. L., HAY, R., and TWOMBLY, J. (1976), *Detection of Hydraulic Fracture Orientation and Dimensions in Cased Wells*, J. Pet. Tech., 1116–1124.

RALEIGH, C. B., HEALY, J. H., and BREDEHOEFT, J. D., *Faulting and crustal stress at Rangely, Colorado*. In *Flow and Fracture of Rocks* (American Geophysical Union, Washington D.C. 1972) pp. 275–284.

SALAMON, M. D. G., *Some applications of geomechanical modelling in rockburst and related research*. In Proc. 3rd Int. Symp. *Rockbursts and Seismicity in Mines* (Balkema, Rotterdam, 1993) pp. 297–309.

SATO, T. (1978), *A Note on Body-wave Radiation from Expanding Tension Cracks*, Sci. Rep. Tohoku Univ., Ser.5, Geophysics *25*, 1–10.

SEGALL, P. (1989), *Earthquakes Triggered by Fluid Extraction*, Geology *17*, 942–946.

TALEBI, S. (1998), *A Seismic Model of Casing Failure*, CANMET Report MMSL 98-08(CR), 49 pp.

TALEBI, S., BOONE, T. J., and EASTWOOD, J. E. (1998a), *Injection-induced Microseismicity in Colorado Shales*, Pure appl. geophys. *153*, 95–111.

TALEBI, S., NECHTSCHEIN, S., and BOONE, T. J., (1998b), *Seismicity and Casing Failures due to Steam Stimulations in Oil Sands*, Pure appl. geophys. *153*, 219–233.

TALEBI, S., YOUNG, R. P., VANDAMME, L., and McGAUGHEY, W. J., *Microseismic mapping of a hydraulic fracture*. In Proc. 32nd *U.S. Rock Mech. Symp.* (Balkema, Rotterdam 1991) pp. 461–470.

VITTORATOS, E., SCOTT, G. R., and BEATTIE, C. I. (1990), *Cold Lake Cyclic Steam Stimulation: A Multi-well Process*, SPE Reservoir Engineering, 19–24.

WALTER, W. R., and BRUNE, J. N. (1993), *Spectra of Seismic Radiation from a Tensile Crack*, J. Geophys. Res. *98*, 4449–4459.

(Received July 2, 1998, accepted September 25, 1998)

Pure appl. geophys. 153 (1998) 219–233
0033–4553/98/010219–15 $ 1.50 + 0.20/0

│Pure and Applied Geophysics

Seismicity and Casing Failures Due to Steam Stimulation in Oil Sands

S. Talebi,[1] S. Nechtschein[2] and T. J. Boone[2]

Abstract—This paper describes observations of seismicity and casing failures associated with steam stimulation operations at Imperial Oil Ltd.'s Cold Lake oil field in Alberta, Canada. A total of 11 oil-producing pads were monitored over a 1–2 year period using 3-component geophones cemented at depths ranging from 160 m to 400 m and data acquisition systems with a flat frequency response up to 1.5 kHz. Most of the seismicity was detected during the steaming operations and was located in the formation overlying the oil-bearing layer. Some activity was observed in the shales above, however, the reservoir itself showed almost no evidence of seismicity. The estimated seismic moment of the observed events was in the range 10^5–10^7 N·m ($-2.7 < M < -1.3$). According to a theoretical model (TALEBI *et al.*, 1998) and *in situ* observations, the seismic source corresponding to casing failure events should be well described by a dipole registering seismic moment in the order of $2 \cdot 10^6$ N·m. Seismic signals of a total of four observed casing failures were analyzed. The partial failures produced seismic moments slightly lower than this value while total failures were stronger by about one order of magnitude. The use of the SV/SH amplitude ratio, in conjunction with accurate source locations, provided a robust technique for the detection of casing failures.

Key words: Casing failure, oil sands, seismicity, seismic modeling, steam stimulation.

Introduction

The large majority of Canada's oil sands deposits are too deep to be mined from the surface and too viscous to be exploited employing conventional extraction techniques. The main obstacle facing the economic recovery of hydrocarbons from these deposits is their extremely high viscosity; hence thermal recovery processes are commonly used. Cyclic Steam Stimulation (CCS), developed by Imperial Oil Resources Ltd. and used in their Cold Lake oil field in Alberta (Fig. 1), is a technique that uses the same well for steam injection and oil production (BUCKLES, 1979; KRY, 1989; VITTORATOS *et al.*, 1990). Access to the reservoir is achieved by using directional drilling from a central surface location called a pad in order to

[1] CANMET, 1079 Kelly Lake Rd., Sudbury, Ontario, Canada P3E 5P5. Fax: 705-670-6556. E-mail: stalebi@nrcan.gc.ca
[2] Imperial Oil Resources Ltd., 3535 Research Rd. NW, Calgary, Alberta, Canada T2L 2K8.

minimize surface disruptions (Fig. 2). The extraction process consists of injecting large volumes of steam under high pressures and temperatures (300°C) into the oil-bearing Clearwater formation, located at about 450 m of depth, in order to reduce the viscosity of the bitumen and allow its natural flow towards producing wells. Steam injection alternates with oil and water production for as many cycles as economic conditions permit (KRY, 1989). The safe exploitation of the Cold Lake deposit using CSS was possible because of the ambient stress field. It is well established that hydraulic fracturing within the reservoir initiates vertical fractures in a northeast-southwest direction (BUCKLES, 1979). This is due to the fact that the minimum principal stress is horizontal and lies in a northwest-southeast direction (DUSSEAULT, 1980). At depths less than about 400 m, the minimum principal stress corresponds to the overburden and, consequently, hydraulic fractures become horizontal (DUSSEAULT, 1977). Therefore, massive fluid injections within the reservoir produce hydraulic fractures that ultimately remain confined within the reservoir itself.

One possible operational problem of using CSS at the Cold Lake deposit is the failure of well casings at the level of the Colorado Shale formation which extends roughly from 150 m to 320 m of depth; the 200–250 m depth range is of particular concern. The development of a reliable tool capable of early detection of

Figure 1
Map of western Canada showing the location of the Cold Lake Deposit.

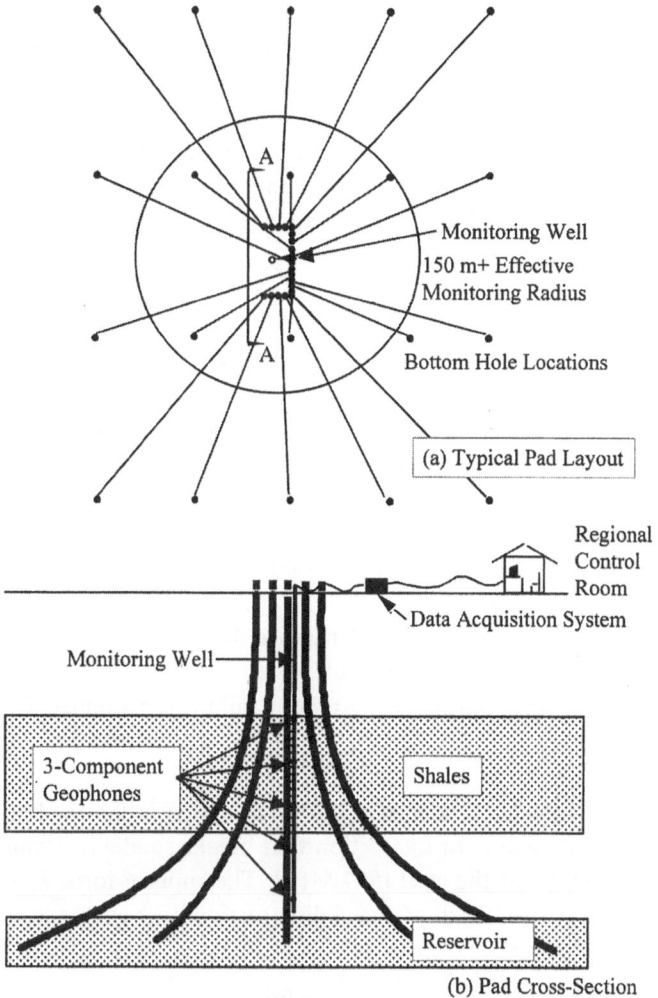

Figure 2
Plan view (a) and cross section (b) of typical pad layout.

such failures is of importance to this operation. Observations of casing failures indicate that, in many instances, such failures are initiated under normal loading conditions as sub-critical crack growth takes place gradually in the steel due to fatigue and stress corrosion, as suggested by BROEK (1986). Brittle rupture of a well casing can occur some time after the completion of a steam injection cycle when boreholes are shut in and well casings are under tensional stresses due to high internal fluid pressure. A declining temperature and the associated fragile behavior of steel further facilitates this process.

Until recently, pressure monitoring of open holes within the Colorado Shale formation was the only technique available for casing failure detection. The method has the advantage of being relatively unintrusive and providing continuous monitoring; but it has serious limitations in that no borehole-specific information is provided since monitoring is performed along, and limited to, a vertical line in the rock mass. Consequently, the results of this method are usually informative but not as reliable as required. Seismic monitoring techniques have the same advantages but allow for monitoring a volume of rock containing all the wells within an oil pad. Proper use of these techniques should allow for identification of specific single or multiple casing failure occurrences.

Theory

The initial concept of casing failure detection was based on monitoring the seismicity induced by fluid percolation through the rock mass, following an episode of casing failure. The detection of casing failures, themselves, requires estimates of the characteristics of expected seismic events due to such occurrences. The seismic model developed by TALEBI *et al.* (1998) was used to estimate the parameters associated with these phenomena. All the equations used in this section are taken from this reference unless specified otherwise. Reasonable numerical values of each parameter are given in square brackets.

The tensional force within the casing at failure is estimated from:

$$F_c = \pi d B f r \sigma_y, \tag{1}$$

where d is the diameter [7 inches = 17.78 cm] and B is the thickness of the casing [10 mm], fr is the fraction of the section that breaks suddenly [estimated at 2/3] and σ_y is the yield stress of the steel [550 MPa]. The rupture force F_c is then estimated to be 205 tonnes. The length of the sliding section L at each end of a failed casing can be obtained from:

$$L = \frac{2 B f r \sigma_y}{\mu_s \sigma_n (2 - \gamma)}, \tag{2}$$

where μ_s is the static coefficient of friction between the casing and the rock mass [0.7], σ_n is the normal stress on the outside of the casing [unknown], and γ is a parameter which defines the loss of friction according to the following equation:

$$\gamma = \frac{\mu_s - \mu_d}{\mu_s} = 1 - \frac{\mu_d}{\mu_s}, \tag{3}$$

where μ_d is the dynamic cofficient of friction. This parameter has been estimated to be about 90% to 95% of its static counterpart (JAEGAR and COOK, 1979). Hence, one assumed a maximum value of $\gamma = 0.1$. Estimation of L from Equation (2) requires that the normal stress on the casing be evaluated. Consider the displacement at each end of a failed casing section at the termination of rupture:

$$\delta(o) = \frac{\mu_s \sigma_n L^2}{6BE} (3 - 2\gamma), \tag{4}$$

where E is the Young's modulus of steel [200 GPa]. Replacing L from Equation (2) into Equation (4) results in:

$$\delta(o) = \frac{2Bfr^2\sigma_y^2(3 - 2\gamma)}{3\mu_s \sigma_n E(2 - \gamma)^2}. \tag{5}$$

Past observations of casing breaks in Colorado Shales indicate that this parameter is usually in the range 0–2.5 cm, but can sometimes be as high as 4.5 cm. Taking 2.5 cm as an average value and using Equation (5), the normal stress on the casing is estimated at about 2 bars. Also, using Equation (2), the length L is estimated at 27.7 m. The expected seismic moment can be estimated from:

$$M_0 = \tfrac{1}{3}\pi d\gamma \mu_s \sigma_n L^2. \tag{6}$$

Combining Equations (4) and (6) yields seismic moment as a function of displacement:

$$M_0 = \frac{2\pi d\gamma BE}{3 - 2\gamma} \delta(o). \tag{7}$$

In other words, the seismic moment of the dipole depends linearly on the final displacement of each end of the casing section. Using Equations (6) and (7), the seismic moment is estimated at $2 \cdot 10^6$ N·m. Adopting the moment-magnitude scale (HANKS and KANAMORI, 1979):

$$M = \tfrac{2}{3}\log M_0 - 6, \tag{8}$$

where M_0 is in N·m results in a moment magnitude of -1.8. Microseismic monitoring during the pilot project resulted in events in the moment magnitude range -3 to -1 (TALEBI and BOONE, 1998). Since the expected average casing failure event falls within this range, it was concluded that the basic design of the seismic equipment used in the pilot project should allow for adequate detection of such occurrences. Now consider the change in the stress field at the completion of the casing failure episode. In the horizontal plane perpendicular to the borehole, the two non-zero components of the change in the stress field, in cylindrical coordinates, can be expressed as follows:

$$\sigma_{rr} = \frac{Mm}{8\pi\mu r^3} [2\lambda + (4\mu - \lambda)\Gamma], \tag{9}$$

$$\sigma_{zz} = \frac{Mm}{8\pi\mu r^3} [2(\lambda + 2\mu) - \Gamma(\lambda + 6\mu)], \tag{10}$$

where Mm is a moment defined by the product of the length L and the force applied to the outside of the casing (equal to F_c), λ and μ are Lamé constants for the shale and r is the horizontal distance from the failed section. Γ is estimated at 8/9 using a value of 3 for the ratio of P- to S-wave velocity for shales (TALEBI *et al.*, 1998). Using a ratio of $\lambda/\mu = 7$, obtained from the ratio of P/S-wave velocity equal to 3, combined with estimates of other parameters, Equations (9) and (10) can be rewritten as follows:

$$\sigma_{rr} = 25.6/r^3, \tag{11}$$

$$\sigma_{zz} = 14.6/r^3, \tag{12}$$

where stresses are expressed in MPa. Obviously these stress changes are significant close to the area of the casing failure but fade away very rapidly as a function of horizontal distance. Consequently, these stress changes constitute a secondary effect. The main hazard associated with casing failure occurrences is related to the leakage of fluids in the formation.

The Equipment

A total of eleven oil-producing pads were equipped with microseismic sensors and data acquisition systems spanning two years. Because of the constraint that applications of this magnitude should be based on using only one borehole per pad for sensor installation, each pad was monitored using five 3-component geophones installed in a vertical well. These observation wells were centrally located at each pad, no closer than 10 m from any producing wells, and extend to the top of the Clearwater formation (Fig. 2). Sensors were equally spaced between the top of the Colorado Shales and the top of the Clearwater formation (i.e., depths of about 160 m, 220 m, 280 m, 340 m and 400 m). Coupling of the sensors to the rock mass was achieved by permanently cementing the geophones in place.

Preamplifiers were located in appropriate housings, mounted on a post at the well heads. The A/D board of a data acquisition system at each pad detects the activity on the 15 available channels and records microseismic events when user-defined criteria on signal amplitudes on a sufficient number of channels are met. This system uses a QNX operating system, which is a PC-based version of UNIX designed for real-time monitoring. Calibration tests have shown a flat frequency response for the whole system up to 1.5 kHz. The sampling rate is 3 kHz per channel. The system is accessed remotely from Calgary on a daily basis and data are downloaded for processing via a phone line.

Figure 3 displays typical P-wave signals recorded on five 3-component sensors following primacord shots detonated in shallow surface holes. The horizontal shot-sensor distance for this example was about 600 m. The orientation of horizontal components of geophones was estimated using P-wave polarization analysis (e.g., MATSUMURA, 1981). Note the high quality of raw and rotated signals, indicating that a perfect coupling of the sensors to the ground has been achieved. The standard deviation of orientation estimations for 4–5 shots at various locations relative to each sensor array was typically in the range 1–2 degrees. This indicates a high degree of confidence in the azimuth information from the sensors. Note that all first motions of P waves on rotated signals are upwards, indicative of a compressional source.

Seismic Activity

In 1996, seismic monitoring was initiated at four oil-producing pads undergoing their fourth steaming cycle and continued over a 2-year period. The small number of seismic events recorded indicated that very little seismicity could be expected for

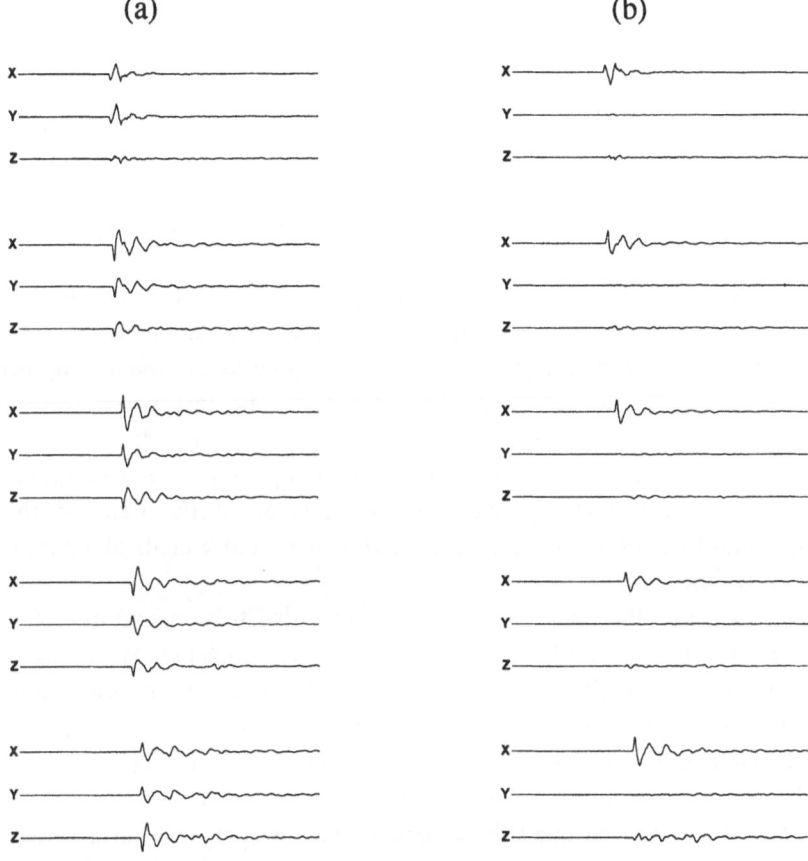

Figure 3
(a) Raw and (b) rotated P-wave signals of a primacord shot detonated on surface. The time window is 500 ms.

(a)   (b)

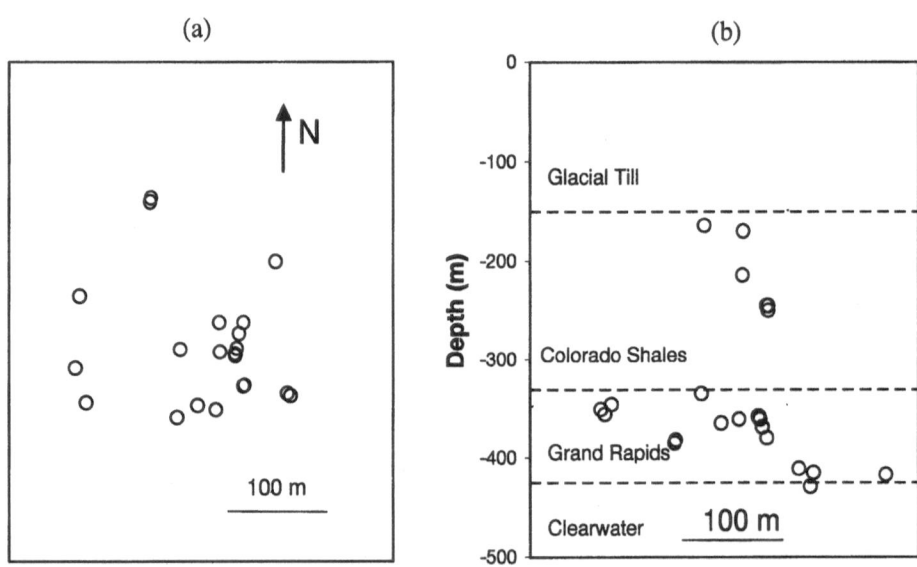

Figure 4
(a) Plan view and (b) vertical section looking north of seismicity of an oil pad.

such pads that are in an advanced stage of their exploitation. In 1997, seismic monitoring was extended to an additional 7 pads and continued for one year. Three of these pads were undergoing their fourth steaming cycle too and exhibited relatively low levels of seismicity as well. Seismicity of another 3 pads that were being steamed for the first time was monitored during steaming cycles 1–3. Most of the seismicity was detected during the first steaming cycle and the activity declined significantly during the second and third cycles. The last pad where seismic monitoring was performed was not subjected to steam injection. Interestingly, some seismic events were still recorded due to steaming operations on an adjacent pad. Observation of lateral casing deformations in some of the wells of this pad indicated that these events were associated with shear movements along a plane of weakness within the rock mass.

Figure 4 shows the microseismic activity recorded from a pad during its first steaming cycle that took about two months to be completed. We focus on this particular pad since it provided pertinent observations of events originating from within the rock mass, as well as those caused by sudden failure of well casings. The notable result is the large number of microseismic events detected in the Grand Rapids formation overlying the deposit. The same was the case as well for the two other pads that were steamed for the first time. Activity was also observed in Colorado Shales depending on the pad and stage of exploitation. However, the Clearwater formation, which was the target of steam stimulation, shows almost no evidence of seismicity, even during early cycles. Numerical models indicate that injection of large volumes of steam under high pressures and temperatures causes this formation to expand, which increases shear stresses on the overlying formations. The observed seismicity can then be explained by the increase in shear stresses because of thermal expansion of the deposit. Another possible explanation is that seismicity originates from the decrease in normal effective stresses due to the percolation of fluids within the Grand Rapids formation. The two explanations are equally acceptable and a combination of the two effects cannot be excluded.

Figure 5 shows signals of a seismic event located within the Grand Rapids formation on four 3-component geophones at depths 220 m, 280 m, 340 m and 400 m. The signals are very clean and show clear arrivals of P and S waves with no significant secondary or coda components. The rotated signals reveal the presence of both SH and SV components with comparable amplitudes (Y and Z traces, respectively). This is usually the case for most recorded events that display, occasionally, saturated S-wave signals. These events seem to be caused by shear failure along natural fractures or contact planes between different layers within the rock mass. Evidence of the presence of the latter type of events becomes clear as they tend to cluster on some known contact planes. The seismic moment of recorded events was typically in the range 10^5-10^7 N·m (moment magnitudes in the range -2.7 to -1.3).

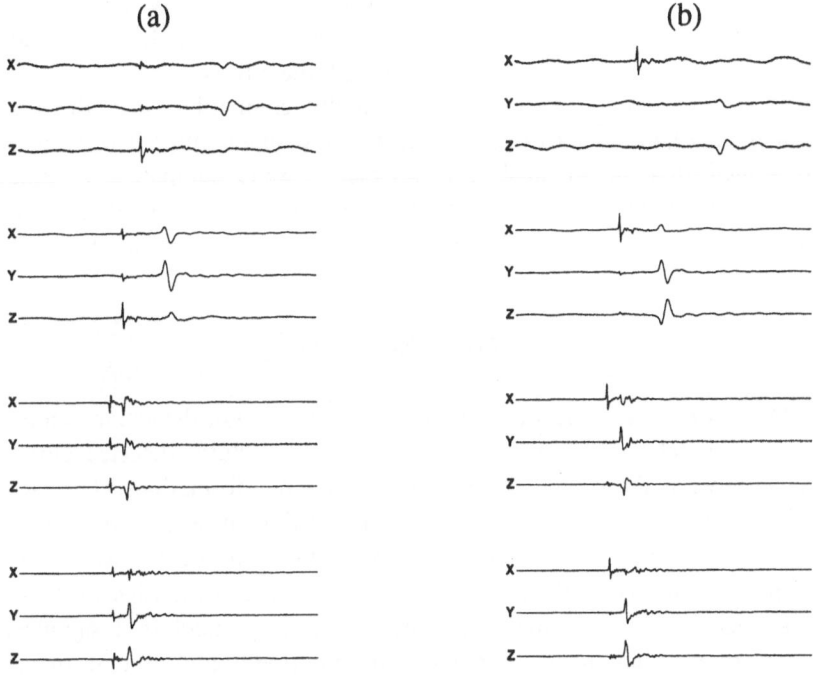

Figure 5
(a) Raw and (b) rotated P-wave signals of a microseismic event. The time window is 500 ms.

Casing Failures

Analysis of seismicity recorded during the first steaming cycle of the same pad revealed three microseismic events located on one of the oil-producing wells at depths between 244 and 246 m. The first two events occurred during the first day of steaming while the third event was recorded two days after the end of the operation. Later, all the wells of this pad were purged with nitrogen and a leak was discovered at a depth of 247 m in the same borehole in which three events were detected. A camera was then run inside the borehole and a damaged connection was observed at this same depth with a displacement of about 1.25 cm at each end of the casing. From Equation (7), such a displacement would be expected from a dipole with a seismic moment equal to 10^6 N·m. Full technical assessment of the situation, based on *in situ* data and visual observations, indicated that these events were caused by three sudden jumps of threads of the connection at 247 m of depth. The most likely scenario is that the first two events were caused by the expansion of the well casing which was subjected to high temperature increases at the beginning of steaming. The third event, however, was caused by the shrinkage of the same casing which was cooling off at the end of the operation. Interestingly, analysis of first motion of P waves corroborates this interpretation in that clear dilatational first motions were observed for all the rotated signals of the first two events while the opposite was the case for the third event (Fig. 6). The source of these events is then compatible with a dipole that flips its polarity depending on whether the casing expands or shrinks. Although these three events do not represent complete casing failures, they are conceptually compatible with the model of TALEBI *et al.* (1998) if one treats them as partial casing failures. The whole episode demonstrated that, as planned, the designed seismic equipment is capable of detecting casing failure and that the accuracy of source location determinations for the three events was within 2–3 m of the observed damage location.

Processing Results

Table 1 shows the results of signal processing for two of the above events as well as two additional events generated by total casing failures detected close to the interface between the Grand Rapids and Clearwater formations. The latter two events were recorded 1–3 days after the completion of steam injection and are tensional casing failures. The accuracy of source location of these events was in the order of 20–30 m due to the fact that they are located at the edge of the array of sensors. Signals from all four events evidenced very small to insignificant *SH* compared to *SV* components after signal rotation (see Fig. 6), in agreement with a dipole source. Likewise, the *SV/SH* amplitude ratio method, in conjunction with accurate source locations, provided a robust technique for casing failure detection,

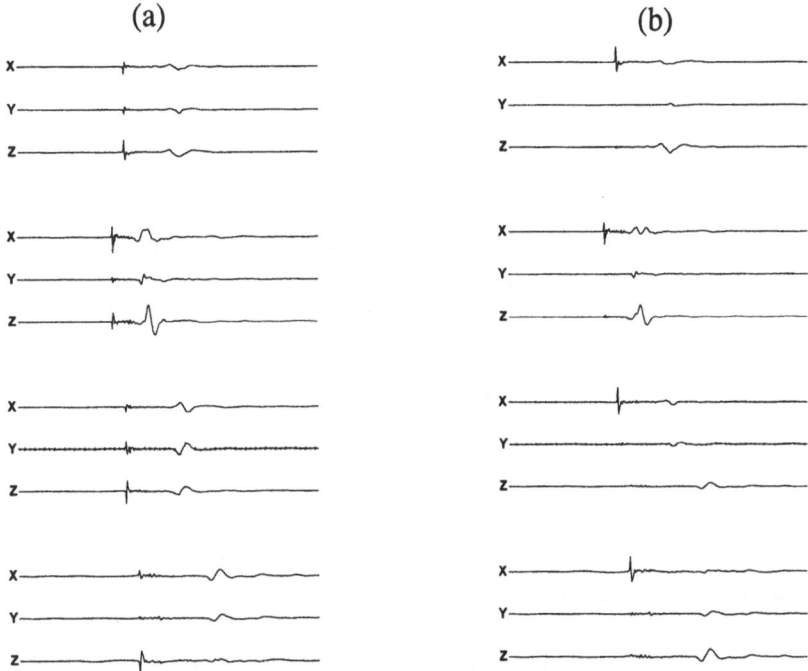

Figure 6
(a) Raw and (b) rotated signals from a partial casing failure (event 2) on four 3-component geophones located at depths of 160 m, 280 m, 340 m and 400 m. The time window is 500 ms.

as postulated by TALEBI et al. (1998). Seismic moment for these casing failure events was obtained from P-wave signals in time and frequency domains. In the time domain, the following equation was used (same authors):

$$M_0 = \frac{4\pi\rho\alpha^3 r}{R^P} \int U_r(r, t) \, dt, \tag{13}$$

where ρ is the density of the source material [2100 kg/m^3], α is the P-wave velocity [2000 m/s], r is the source-sensor distance and R^P is the P-wave radiation pattern coefficient. The integration is performed over a suitable time period containing the P-wave displacement pulse. Radiation pattern coefficients were calculated from (same authors):

$$R^P = \cos^2 \theta, \tag{14}$$

where θ is the angle between the direction of the source dipole and source-sensor orientation. In the frequency domain, displacement spectra of the rotated P-wave signals were calculated and the following equation was used to provide a fit to these data, following the methodology of TALEBI and BOONE (1998);

$$U(\omega) = \frac{\Omega_0 \, e^{-\omega r/2\alpha Q_P}}{1 + \left(\dfrac{\omega}{\omega_0}\right)^2},$$

(15)

where Ω_0 is the level of the low-frequency trend of displacement spectra, α is the P-wave velocity, Q_P is the P-wave quality factor, ω and ω_0 correspond, respectively, to the angular frequency and the angular corner frequency. As shown in Figure 7, this procedure provided a satisfactory fit to displacement spectra and estimates of three parameters (Ω_0, ω_0 and Q_P) were made. The P-wave quality factors are somewhat higher than expected for the two shallow events. Lower values were obtained for the two deeper events. The seismic moment was then obtained from:

$$M_0 = \frac{4\pi\rho\alpha^3 r\Omega_0}{R^P}.$$

(16)

We now examine the results of Table 1 in detail. First, it is worth mentioning that the high gain levels used early in the monitoring, particularly for the sensors in the Colorado Shales, caused signal saturation in some cases; and, therefore, less than five determinations were possible. Except for the results of the first sensor for the second event, seismic moment estimates for each event obtained from different sensors are reasonably comparable to each other. Also, seismic moment estimates from time and frequency domains are reasonably close in most cases. This clearly indicates that both methods are equally valid for seismic moment determinations. However, contrary to the expectation, some of the time-domain results are slightly

Table 1

Results of signals processing of four casing failure events: R^P is radiation pattern coefficient, M_0 is the seismic moment calculated in time and frequency domains, Q_P is the quality factor of P waves and M is moment magnitude

Event	Depth (m)	Mechamism	Sensor depth (m)	R^P	M_0 (time) (10^6 N·m)	M_0 (freq.) (10^6 N·m)	Q_P	M
1	247	Partial failure	341	0.79	1.6	1.4	58	−1.9
		Compressional	398	0.90	2.6	1.6	75	−1.9
2	247	Partial failure	155	0.76	2.5	1.4	35	−1.9
		Tensional	280	0.33	0.6	0.6	40	−2.1
			341	0.79	0.5	0.6	50	−2.1
			398	0.90	0.4	0.6	45	−2.1
3	438	Total failure	155	0.93	15.7	9.3	33	−1.4
		Tensional	213	0.97	9.1	8.2	35	−1.4
			398	0.77	16.3	15.2	8	−1.2
4	430	Total failure	160	1.00	28.1	32.2	17	−1.0
		Tensional	222	0.98	27.6	30.2	14	−1.0
			285	0.91	30.9	38.9	11	−0.9
			347	0.75	33.8	49.6	8	−0.9
			406	0.54	41.7	48.9	13	−0.9

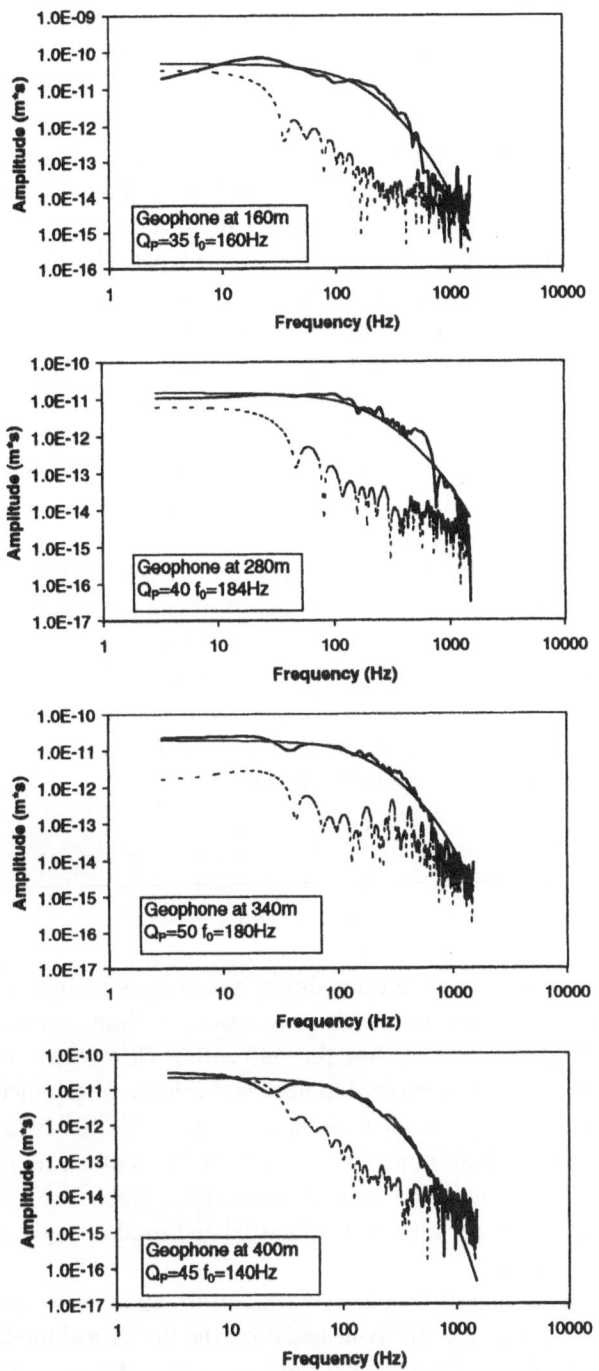

Figure 7
Displacement spectra of *P*-wave signals from the casing failure event 2 recorded on four 3-component geophones. The best fit to the spectra using Equation (15) is also shown. The spectrum of background noise is shown by dotted lines.

larger than their frequency-domain counterparts and constitute overestimations of the seismic moment. This is due to the presence of a low-frequency noise component in some signals. Partial failures have seismic moments slightly lower than the predicted value of $2 \cdot 10^6$ N·m while total failures are stronger by about one order of magnitude. Moment magnitudes are, respectively, about -2 and -1 for the two cases. The larger magnitude of deeper events is probably associated with larger post-failure displacements, but field observations have as yet not been attempted. As for the observed displacements in the case of partial failures, the available information cannot be attributed to a single event. It implies, however, events of the order of magnitude of that which has been detected.

The overall results presented in this paper indicate that the model of TALEBI *et al.* (1998) describes reasonably well the observed casing failure occurrences. Prior to this experiment, a contingency plan was contemplated in order to actually break a well casing by filling it with liquid nitrogen so as to record and analyze its seismic signals for an in-depth analysis. The fact that seismic monitoring helped identify total and even partial casing failures eliminated the need for this undertaking. Future work will involve the design of a procedure, based on the above model and past observations, which will attempt to detect casing failures in real time. Also, measurements of displacements at the two ends of failed casings will be attempted, if possible, in order to better understand the interaction between the casings and the rock mass. In the long term, the use of seismic information is expected to further improve the understanding of the behavior of different layers in response to episodes of steam stimulation and oil extraction.

Conclusion

Most of the seismicity was detected during steaming operations and was located in the Grand Rapids formation overlying the reservoir. Some activity was observed in the Colorado Shales as well, but the oil-bearing Clearwater formation which underwent steam stimulation showed almost no evidence of seismicity. The seismic moments of the observed events were in the range 10^5–10^7 N·m ($-2.7 < M < -1.3$). Theoretical considerations and past observations of casing failure indicated that such occurrences should be well described by a dipole source. Processing of seismic signals of a total of four observed partial and total casing failures confirmed the validity of this model.

Casing failures within the Colorado Shale formation were expected to have seismic moments of about $2 \cdot 10^6$ N·m based on the theoretical model of TALEBI *et al.* (1998). Observations of four casing failures revealed seismic moments in the order of 10^6 N·m for partial failures within the same formation. Total casing failures at larger depths were larger by at least one order of magnitude. The use of

SV/SH amplitude ratio and accurate source locations provided a robust technique for the detection of casing failures. This can contribute to the safe and economical operation of the Cold Lake oil field.

Acknowledgments

The authors would like to thank Imperial Oil Limited, Resources Division, for their financial support of this project and permission to publish the data. Comments from an anonymous reviewer helped in improving the quality of this paper.

REFERENCES

BUCKLES, R. S. (1979), *Steam stimulation heavy oil recovery at Cold Lake, Alberta*. In *Proc. Calif. Reg. Meeting the SPE of AIME*, Ventura, Calif., April 18–20, SPE 7994, 1–12.

BROEK, D., *Elementary Engineering Fracture Mechanics* (Martinus Nijhoff, Dordrecht, The Netherlands 1986).

DUSSEAULT, M. B. (1977), *Stress State and Hydraulic Fracturing in the Athabasca Oil Sands*, J. Can. Pet. Techn. *16*, 19–27.

DUSSEAULT, M. B. (1980), *The behaviour of hydraulically induced fractures in oil sands*. In *Proc. 13th Canadian Rock Mech. Symp.*, pp. 36–41.

JAEGAR, J. C., and COOK, N. G. W., *Fundamentals of Rock Mechanics* (Chapman and Hall, London 1979).

HANKS, T. C., and KANAMORI, H. (1979), *A Moment Magnitude Scale*, J. Geophys. Res. *84*, 2348–2350.

KRY, P. R. (1989), *Field observations of steam distribution during injection to the Cold Lake reservoir*. In *Proc. Symp. Rock at Great Depth*, Pau, France, pp. 853–861.

MATSUMURA, S. (1981), *Three-dimensional Expression of Seismic Particle Motions by the Trajectory Ellipsoid and its Application to the Seismic Data Observed in the Kanto District, Japan*, J. Phys. Earth *29*, 221–239.

TALEBI, S., and BOONE, T. J. (1998), *Source Parameters of Injection-induced Microseismicity*, Pure appl. geophys. *153*, 113–130.

TALEBI, S., BOONE, T. J., and NECHTSCHEIN, S. (1998), *A Seismic Model of Casing Failure in Oil Fields*, Pure appl. geophys. *153*, 197–217.

VITTORATOS, E., SCOTT, G. R., and BEATTIE, C. I. (1991), *Cold Lake Cyclic Steam Stimulation: A Multi-well Process*, SPE Reservoir Engin., 19–24.

(Received July 12, 1998, accepted September 25, 1998)